"Anyone ready to innovate outside the box will be blown away by the vision and practical insights demonstrated in *Growing Hybrid Hazelnuts*. The eclectic blend of science and practical how-to information packed into this fascinating, readable book is enough to inspire a whole new generation of farmers. Turning soybean fields into hazelnut plantations is truly a vision for the stout-hearted pioneer futuristic farmer. Sign me up."

—**Joel Salatin**, farmer and author of *Fields of Farmers*

"The authors of *Growing Hybrid Hazelnuts* have been advocating woody-plant staple crops for carbon sequestration since I was in high school. This manual provides theory, context, budgets, and practical details like pest management and fertility for this important new crop. Worth the price for the information on their innovative hybrid swarm breeding system alone."

—**Eric Toensmeier**, author of *Paradise Lot* and coauthor of *Edible Forest Gardens*

"This book is not only a testament to forty years of dedicated hazel-breeding work, but also a call for more tree crops for a sustainable agriculture in general—a fantastic manual about all aspects of cultivating hardy hazel trees and processing their nuts."

—**Martin Crawford**, author of *Creating a Forest Garden*

"If you are dissatisfied with the current state of the annual-based agriculture system currently dominated by corn and soy, then here is an alternative. This book is an action plan building off of Phil Rutter's thirty years of experience, where you can become part of the actual on-the-ground change toward building a perennial woody system that conserves resources while providing for human needs."

—**Diego Footer**, founder of Permaculture Voices

"*Growing Hybrid Hazelnuts* is a compelling work combining natural history, genetics, and ecology to form a rich strategy for breeding hardy, disease-resistant, and productive hazelnuts. The need for perennial staple crops is great, and the authors show that it takes time, integrity, and patience to develop a crop that will feed the world. This book not only completely covers hybrid hazelnut cultivation, it also offers a roadmap for breeding other crops if we are to get serious about regenerative perennial food production."

—**Steve Gabriel**, coauthor of *Farming the Woods*

"A more resilient future requires diverse and hardy food-bearing crops. *Growing Hybrid Hazelnuts* is an encyclopedia of the 'how-to' and 'why-for' of breeding, growing, harvesting, and marketing this unique and important crop. We just planted 300 hazelnuts last year!"

—**Nathan John Hagens**, president of Bottleneck Foundation

Growing Hybrid Hazelnuts

*The New Resilient
Crop for a Changing Climate*

Philip Rutter, Susan Wiegrefe, and
Brandon Rutter-Daywater

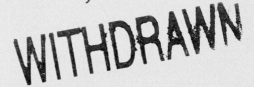

Chelsea Green Publishing
White River Junction, Vermont

Acquiring Editor: Makenna Goodman
Developmental Editor: Fern Marshall Bradley
Copy Editor: Laura Jorstad
Proofreader: Eric Raetz
Indexer: Shana Milkie
Designer: Melissa Jacobson
Page Composition: Abrah Griggs

Botanical illustrations of the Badgersett neohybrid hazelnut
by Denise Walser-Kolar

Printed in the United States of America.
First printing January, 2015.
10 9 8 7 6 5 4 3 2 1   15 16 17 18

Our Commitment to Green Publishing
Chelsea Green sees publishing as a tool for cultural change and ecological stewardship. We strive to align our book manufacturing practices with our editorial mission and to reduce the impact of our business enterprise on the environment. We print our books and catalogs on chlorine-free recycled paper, using vegetable-based inks whenever possible. This book may cost slightly more because it was printed on paper that contains recycled fiber, and we hope you'll agree that it's worth it. Chelsea Green is a member of the Green Press Initiative (www.green pressinitiative.org), a nonprofit coalition of publishers, manufacturers, and authors working to protect the world's endangered forests and conserve natural resources. *Growing Hybrid Hazelnuts* was printed on paper supplied by QuadGraphics that contains at least 10% postconsumer recycled fiber.

Library of Congress Cataloging-in-Publication Data
Rutter, Philip, 1948- author.
  Growing hybrid hazelnuts : the new resilient crop for a changing
climate / Philip Rutter, Susan Wiegrefe, and Brandon Rutter-Daywater.
    pages cm
Includes bibliographical references and index.
  ISBN 978-1-60358-534-7 (pbk.) — ISBN 978-1-60358-535-4 (ebook)
1. Hazelnuts. I. Wiegrefe, Susan, author. II. Rutter-Daywater,
Brandon, author. III. Title.
  SF401.F5R88 2015
  634'.5--dc23
                    2014039561

Chelsea Green Publishing
85 North Main Street, Suite 120
White River Junction, VT 05001
(802) 295-6300
www.chelseagreen.com

MIX
Paper from
responsible sources
FSC® C084269

For the farmers, those they feed,
and the children. All of them.

Contents

Introduction   ix

1. Hazels, Hybrid Hazels, and Neohybrid Hazels   1
2. The Mortal Sins of Modern Agriculture   23
3. The World Hazelnut Industry   39
4. Preparing to Plant   47
5. Planting and the Establishment Period   87
6. Pest Management   105
7. Managing Mature Hazels   135
8. Harvest   151
9. Processing: From Harvest to Market   163
10. Marketing   177
11. Co-Products and Their Value   187
12. Neohybrid Hazels—Beyond Mendel   197
13. Climate Change, Resilience, and Neohybrid Hazels   211
14. The State of the Crop   225

Acknowledgments   233
Notes   234
Resources   241
Index   248

Introduction

Growing neohybrid hazels is a radically new way to produce the food our species needs. It's also a complex topic to present and explain. One realization we came to while writing this book is that we do not think of the problems in world food systems as linear or even three-dimensional. We see the problems and our possible solutions as five-dimensional, at least. So where do we start to explain?

As you read this book, please keep two things in mind. First: We're discussing systems that interact in interrelated cycles, over the course of many years. If what you're reading doesn't quite make sense to you right away, be patient and keep on reading. It's very possible your understanding will blossom as all the pieces become connected into the whole. Please give it a chance.

Second: When a piano student is presented with a new piece of music to learn that is one level of difficulty higher, only 1 student in 10,000 can play that piece through on the first try. The rest have to fight their way through the music in jerks and stops, and frequently with muttered "I'll never get this!" complaints. But in two weeks, with time and practice, most will be playing it by heart. This book may work that way, too.

If you are reading this book, chances are you are already deeply familiar with the environmental problems caused by modern agriculture. Modern crop scientists working with rice, maize, and so on are busy looking for ways to adapt their crops to meet future needs, most frequently via genetic engineering. Yet independent analysts and statisticians almost without exception come to the conclusion that nothing they are working on will be enough. Not even close. What we must have, the analysts say, is new crops, new systems, that work in some unspecified but totally different fashion, so that we do not destroy our environment.[1]

We are almost 40 years into the creation of a crop that might meet all these needs: "neohybrid" hazelnuts. It is ready for widespread and scaled-up plantings—a true perennial agriculture, based on woody plants.

But. The requirements for creating a widely adapted, diverse, and adaptable crop include the need for many plantings, in many places and conditions, over many years, and the focused attention of many, many eyes. This history of large-scale, decades-long projects conducted by universities is totally uniform—they decline, then disappear, a few years after the founder retires or passes away. Universities are important, but historically they follow farmers' innovations; they do not lead. Our modern US soybean crop was created by thousands of farmers, working together. Their vision, their sweat, their risks, their perseverance. That is what we must have again, now. Your eyes and your children's. With the genetics we have already developed, you can soon be growing food, fuel, and more for your family. New doors are opening rapidly.

Are you ready?

CHAPTER 1

Hazels, Hybrid Hazels, and Neohybrid Hazels

How do you create a crop?

The world is in urgent need of an answer to that question.

An abundance of statistics shows that climate change is already reducing world crop yields.[1] And while croplands are declining in productivity worldwide—and analysts expect the decline to continue—another world food catastrophe looms as well. Not only is our current food production decreasing steadily due to climate change, but because of population growth we need to *increase* food production dramatically. In 2009, the United Nations Food and Agriculture Organization (FAO) issued this statement of its best calculation of crop production and food needs: "The projections show that feeding a world population of 9.1 billion people in 2050 would require raising overall food production by some 70 percent between 2005/07 and 2050."[2]

There are, simply, no technologies in sight that can cope with these problems. And in addition to climate change and population growth, global political and economic stability seem to develop new crises daily, which makes coping with change more difficult, adaptation more chancy.

From any perspective, the world is in desperate need of new crops and new food production practices. Crops able to feed humanity despite unpredictable and more extreme weather, the appearance of new pests as local climates shift, and other uncertainties we cannot even imagine or predict. Crops able to produce what we think of as Food, with the capital *F*, able to feed the cities where more than half of all humanity now lives.

What would such crops look like? You will hear expert opinions about that for the rest of your life. And loud claims from entrepreneurs, corporations, governments, and charlatans. They're already cashing in on the fogs of doubt and fear.

The wise people of all cultures of the world have offered this advice: Learn from your past. So let's start by asking: How were new food crops created in the past? The beginnings of many of our common food crops have now been well studied. We've learned that all the plants we depend on for food had their origins in the common people: unknown ancient farmers who grew crops to feed their families and their villages. Thousands of farmers, slowly refining and selecting crops over thousands of years.

The greatest crop breeders the world has ever known were unquestionably the pre-European-invasion Native Americans. While our Old World ancestors took wild grass and made bigger grass (wheat, rye, oats, barley, rice), our New World ancestors took a wild grass called teosinte and transformed it into maize. One ear of maize contains not double, or triple, the amount of food in a seed head of teosinte—but closer to 50 times the amount. Indeed, maize is so vastly more productive than any wild grass, and so different in its fruit, the corncob, that until 2004 and easier DNA analysis[3] many top scientists argued it could not possibly have been derived from teosinte. Genetic fingerprinting proved it was, and that it had been bred, selected, and created by pre-Columbian Native Americans. Those peoples also gave us potatoes, pumpkins, squash, all beans but soy and fava, turkeys, tomatoes, peppers, amaranth, quinoa, and more.

They did it as villages, and it wasn't just the medicine men who were interested in the crops: Entire villages harvested and selected; found the best every year, and carefully planted it the next; and cheered the changes and improvements.

By contrast, modern experts—government projects, corporations, entrepreneurs—do not create truly "new" crops. They have made considerable and important incremental improvements in the crops and farm systems that our ancestors gave us. In no case, however, have modern experts duplicated the astonishing achievements of the original common people.

Do you think the various governments, non-governmental agencies, and philanthropists of the world will address the looming food crisis effectively? It's not a matter of *whether* they can; the power and resources probably exist. But *will* they?

I don't see much chance of that happening. Over the course of writing this book, I've seen my estimate of any such chance drop several percentage points. That's why I'm working with neohybrid hazelnut production, and why I've written this book.

Neohybrid hazelnuts are a genuinely new kind of crop requiring an entirely new kind of growing system. Our ancestors could not have adopted this path, but we have tools and knowledge they did not. This can be a successful new food and commodity. But it cannot be developed by a university. It cannot be grown by governments. It will take common folks, following their own paths—and working together. Farmers and the common people used to think—and act—for themselves. We need to start doing that again.

Hazelnuts as a New Staple Crop

Imagine for a moment what the environmental benefits would be if we could grow our staple foods—corn, beans, rice, wheat—without plowing. Now take another moment, and imagine it some more.

Perennial crops have been a perennial aspiration, even extending to corn. When corn's wild grass relative *Zea diploperennis* was discovered a couple of decades ago, plenty of mainstream agronomists began to fantasize about, and work on, developing perennial corn. It's a dream that has proven very difficult to realize.

Hazelnuts, by contrast, are a perennial food crop that already exists and that once did serve as an important, storable, staple food for many pre-farming cultures across Asia, Europe, and North America. They could again.

Once established, a planting of hazels requires no plowing or even cultivation. No water runs off the fields, because infiltration rates are dramatically improved regardless of soil type. Tiling should not be necessary in moderately wet soils. No fertilizer ever escapes into groundwater, because the crop has extensive permanent root systems, at work 365 days a year. No soil is lost to wind or rain; in fact, this crop builds soil. Wildlife finds cover and food all year, instead of naked soil for eight months and a monoculture for the other four.

Hazelnuts have a large and expanding world market. The existing world hazelnut crop is based on tree-type hazels from Europe in a system little changed, and with genetics little changed, for hundreds of years. New plantings tend to follow that model, although many plantings in Turkey and Eastern Europe rely on the ancient non-mechanized large bush culture that is tens of thousands of years old. (Turn to chapter 3 to learn more about world hazelnut production.)

 ## Choosing Hazels as a Focus

I first became seriously interested in hazels because I couldn't find any. In the 1970s, my first wife, Mary Lewis, and I both decided to leave our nearly completed work in doctoral ecology programs at the University of Minnesota and move to southeastern Minnesota to live on a farm we found ourselves rather accidentally owning. The intention was to "play pioneer" for a few years while we figured out what we "really wanted to do." We built a log cabin and grew a big garden.

We did not disdain our education at all; I'm one of those who believe no bit of education is ever wasted, on anyone. It enriches your life, particularly if you use it. I've long been fascinated by plants, trees, forests, and human interaction with them. Although my PhD program in ecology was through the Zoology Department, you can't be an ecologist and know nothing about plants. So I had plenty of botany and plant community education crammed into me, and found it all fascinating. The playground of the farm astonished us by having a deep ravine—"coulee" in local vernacular—with almost 40 acres of good forest. It had been logged, but the trees had regrown enough that it was time to think about logging it again. However, it was so steep that logging it couldn't be done casually. And because of that discouragingly steep slope, the native plant community was mostly intact. There were enough big sugar maples that we could make maple syrup and sell it. We did, which is how I made the acquaintance of Helen and Scott Nearing. (*The Maple Sugar Book* is the best education I know of on the history of maple sugar—and everything else, for that matter.)

We also found beautiful native orchids. Shooting star, rare out in prairie but sometimes found under limestone cliffs. It was quite natural, and pure fun, to make a list of the native plants on our land. But there were some things missing, including hazels. I looked hard, but there were no hazels. It wasn't hard to guess why some species were missing. We had rented out the 90-some tillable acres to a "regular farmer" neighbor. The rent money paid the taxes on the farm as well as a good chunk of the contract for deed every year. We discovered that our "back hill" (mostly in contours of corn and hay then) didn't have the topsoil on it that it should have, according to the USDA soil survey maps (dating from the 1950s) we'd looked at before buying the place. The map said the back hill had between 12 and 18 inches of black topsoil covering it. But in 1976-ish, a full foot of that black topsoil was gone. At best, the land had 0 to 6 inches left. Most of the hill now looks pale brown if plowed—the color of subsoil.

We were horrified. I knew what this meant. Agriculture was changing this land into desert at tremendous speed. This incredibly rich farmland, which should feed us all forever—some of the best in the world—was being utterly destroyed. And this was not some random faraway global problem; this was right here. My problem. So I thought: *I'm sitting here with this fancy education aimed right at this problem. I need to use it. If nobody ever starts working on it, it will certainly never get better.* It's a bad habit of mine; if there's a problem right in front of me that I have the ability to attack, then I feel it's my responsibility to attack it. (That's also how The American Chestnut Foundation came to exist: luckily for all, Charles Burnham shared my responsibility quirk.)

My neighbors—for hundreds of miles around—were farmers. They needed crops they could grow that wouldn't wreck the land and that would allow them to support their families. So I began to wonder: What used to grow here, before the plow? I once saw a map of Minnesota's original vegetation, based at least partly on the plants used by the original land surveyors as monument trees for property lines and corners. Known as the Marschner Map for the man who spent untold hours putting it together, it now exists in multiple versions, each one carefully bringing the "correct" names of the different vegetation types up to date, so they differ tremendously from one to the next. The version I first saw said most of the land around my farm, and widely common across the region, was "oak-hazel savanna." All the newer versions mention the hazels too.

So I began to ask: Could we grow hazels as a crop here? The answer from various departments at the University of Minnesota was laughter. I was told that hazels couldn't withstand the winters and they would all die of a deadly blight. So why were there hazels all over my original vegetation map?

I was already planting nuts on the farm at this point, an old interest. Chestnuts of various kinds, walnuts, butternuts, and others. And I had found a tiny book called *Growing Nuts in the North* on the shelf in the early hippie food co-op just off campus. This author—Carl Weschcke—had been growing thousands of nut trees in River Falls, Wisconsin, for several decades and was a past president of the Northern Nut Growers Association. About half the book was about hazelnuts. Hybrid hazelnuts, in particular, which he considered one of the most promising types.

I found out that hybrid hazelnuts had been under development in the Upper Midwest since the 1930s, though progress and attention to the work had been sporadic, unofficial, and fragmented. I kept hazelnuts on my list, and I joined the Northern Nut Growers Association. I also went looking for—and found—Weschcke's farm, and his plantings; I regret hugely that he passed away before I could meet him. He died thinking he'd failed with his hazels, and he did not. I got permission to work on his bushes for three years, then selected the seeds that became the foundation of the genetics my co-authors and I are still working with now. He put in 30 years of work that I built on—and it couldn't have been done any faster. Weschcke didn't have

scientific training and he didn't understand the underlying genetics, but he planted out thousands of hybrid hazels and their seedlings (nearly every one of which I examined). This method is known to work. I also began collecting stock developed by other breeders, including Jack Gellatly, George Slate, Cecil Farris, and John Gordon. My work needed some kind of formal business structure, so I created Badgersett Research Corporation. We started planting the first select seeds at Badgersett in 1982. Fairly quickly we had requests, and then demands, for seeds or seedlings from other nut enthusiasts. The word spread, demand grew, and we found Badgersett also becoming a licensed nursery business.

FIGURE 1-1. This improved neohybrid hazel is EFB-resistant, exhibits total cold hardiness, and has added genes for very heavy crop and annual bearing.

Following a decade of initial testing, major new plantings were made of crosses among various breeding lines, and I began a new round of intensive selection and breeding at Badgersett Research Farm. By 2001, it was becoming clear that the developing neohybrids contained the characteristics necessary for the foundation of a genuine hazelnut industry for the northern US Corn Belt and eventually beyond—perhaps far beyond.

 ## What Are Hazelnuts?

Are hazels bushes or trees? Deep-rooted plants? Shallow? Cold-hardy? Descriptions of hazels (the word *hazel* is commonly used for both the nut and the plant, as is *hazelnut*) are abundant, but very often both highly authoritative in tone and simultaneously startlingly unreliable, and for a reason seldom encountered in the world of botany. The plants are so ubiquitous and abundant (in the Northern Hemisphere), and so ancient in human usage, that until quite recently the details of their lives have simply been ignored.

Everyone assumes that everything about these plants must be known already; must have been studied, tested, and codified long ago. The reality is, every hazel region abounds with growers and experts who *do* know their hazels, to the point that they often know highly detailed facts about hazels that are entirely untrue—except in their own region. Only recently have growers and scientists begun to understand that hazels from different regions may be quite different, and the available diversity is very, very large.

So let's ask that question again. Are hazels trees or bushes? Yes, they are.

Hazelnuts (*Corylus* species and hybrids) are plants whose nuts serve as a food source for a wide range of animals as well as people. The genus *Corylus* is at least 40 million years old, has 15 to 20 species worldwide, and is generally considered to consist of three subgroups, two of which are emphatically bushes or "small bushlike trees" and a third that comprises genuine trees sometimes reaching heights of 110 feet. Some of the bushlike trees can be forced into a single-trunk habit by constant pruning, but if left to themselves they will virtually always be multistemmed.

Neohybrid hazels are not all uniform in size. They may be small multistemmed bushes that never grow taller than 8 feet, but may spread out more than 6 feet wide at the base. Or they may be much larger multistemmed bushes that grow as tall as 18 feet. Usually the short types resemble the American hazel form, and the taller resemble the European, with a non-spreading root crown. The root crown of a hazel bush develops new stems constantly. The plants shed old stems as these stems reach 8 to 10 years of age, for those plants resembling the American or beaked hazel ancestor. European-type stems typically live 5 to 10 years longer.

A basic aspect of woody food crops is that the fruit or nut will be produced in a slightly different place on the plant each year—a little higher,

FIGURE 1-2. The line of hazels behind Philip are all half sibs from one female parent. All the same age (about seven years old); the difference in the one outstandingly tall bush is apparent. There may be multiple explanations for this. One is that the tall bush bears very few nuts. In our experience, the most robust bushes bear the fewest nuts.

perhaps, or more spread out. Old wood usually bears less and less over time. In apple, grape, and older hazel fields, this is addressed by pruning. Every plant is pruned every year, for grapes and traditional hazels; at least every other year for apples.

Rather than incessant pruning, which requires skilled labor, neohybrid hazels are selected for one of the outstanding features of hazel's wild ancestors: When you cut them down entirely—or burn them to the ground in a wildfire—they grow back very strongly with fresh, new shoots. The North American Midwest oak savanna is found everywhere along the transitions from eastern deciduous forest to prairies, from Mississippi to northern Minnesota. In Missouri and moving north, hazel becomes increasingly common; the land from mid-Illinois and Iowa northward has often been designated "oak-hazel savanna." These are fire-based biomes; without frequent fire, they become forest. The American hazels that live there must be able to withstand having their tops burned off at any time of year, because prairie fires can be started by lightning or by humans who need to drive game.

Humans have harvested European hazel for fuel and building materials for tens of thousands of years by a method called coppicing, which

FIGURE 1-3. Elly Rutter is standing beside hazels that were cut entirely to the ground one and a half growing seasons before this photo.

means cutting a tree or bush to the ground, harvesting the regrowth for firewood or construction material, and repeating the process. Coppicing results in vigorous new growth. Hazels are known for being very good at this growth habit. The neohybrids vary somewhat in the ability to deal with it; some resent it, and a few do not survive. We require our breeding stock to go through at least one coppice cycle before we make culling decisions.

Multistemmed perennial plants like this often live very long lives, and our observations both in the wild and at Badgersett have led us to the conclusion that these bushes have a natural life expectancy of at least 500 years—possibly over 1,000 (based on reported crown sizes and measured age of other similarly structured species). New stems sprout from the crown only, not from distant roots. The crown itself can spread over time. Crown tissue is different from both the top and the roots of the plant. It usually has a reddish bark, prefers to be just under the soil surface, and naturally gives rise to further crown, top, and root tissue. Stem and root tissue do not do this in hazels. Not only is crown tissue pluripotent, in that it can give rise to both stems and roots, but it also stores resources and appears to make up the "heart" of the plant.

We generally nudge breeding decisions in the direction of more restricted crown growth, though in some practices a wide crown may prove useful. Perhaps a field of hazels with no aisles at all?

Roots form a broad, deep network, with bark a reddish brown on thicker roots (but not so red as crown tissue). Stems have a grayish, mostly smooth bark with coloration and minute patterns that are often distinct between plants, particularly on stems older than one year.

Hazels are monoecious, meaning that plants bear separate male flowers and female flowers; each plant bears both types. Like all the rest of the birch family, hazels are wind-pollinated. Male catkins are formed in late summer and overwinter in a compressed form, opening up in early spring to shed pollen.

Female flowers are small, only showing the red styles poking out from the tips of the flowering buds. Each female flower corresponds to the potential for a cluster of nuts. One of the most powerful ways to improve a plant's crop index—the relative amount of total photosynthesis that is directed into seed production—is to increase the number of seeds in any given cluster of them. This we are successfully doing with the neohybrids, starting with the number of stigmas in one bud.

Within the cluster, each nut is covered by a husk, the fancy biological name for which is *involucre* (IN-vole-ook-er). Neohybrid hazels show many variations on cluster number, arrangement of nuts within the cluster, consistency of nut size and shape within the cluster, and husk shape, length, and thickness.

The husk serves primarily as a defense barrier, and the characteristics of the husks vary from species to species. Those differences can be critical in terms of harvesting methods. American hazels have fleshy husks that totally enclose the nut during summer and open gradually when the nuts are ripe and drying. Husks contain multiple chemical weapons against insects and vertebrate predators, in both the husk flesh and the dense glandular hairs. The nature of the husk has long been cited as a reason why the American hazel, and its hybrids, could never be commercialized. In the hazel industry, the standard method of harvest has always been to deliberately allow the nuts to fall from the trees or bushes, free of their husks entirely, and then sweep the nuts up off the ground. At Badgersett, we've demonstrated that machines will nicely harvest the nuts, husk and all, directly from the neohybrid bushes, and machines will easily remove the husks. We feel this is a tremendous advantage because there is no worry about microbial contamination of the nuts from soil or animal waste. Also, conventional harvesting requires bare soil under the trees so that the machines can sweep up the nuts. In a rainy harvest season, nuts can be lost in mud or left unpicked because the machines cannot travel on muddy, naked soil.

FIGURE 1-4. My hand points to the cut off top of this hazel. We excavated this root system to learn what they truly look like; the hazel survived. To the right you can see both plain roots, the largest being not more than ¾-inch diameter, and the much thicker "storage" root; found only immediately attached to the crown.

FIGURE 1-5. A neohybrid crown that genetically spreads only a limited amount. This trait is highly heritable.

FIGURE 1-6. This widely spreading crown is typical growth for most pure American hazels. As the plant matures, it can make harvest increasingly difficult.

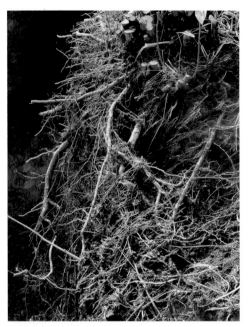

FIGURE 1-7. Everything you can see here is hazel roots. You're looking into a 6-foot-deep trench opened up using a backhoe. Six feet down the roots were still going; so we dug another 5 feet deeper with a posthole digger—the roots were still ⅜-inch diameter and going straight down. We kept the trench open for two years because every plant/tree professional we showed it to was astonished; the root system is far bigger than anyone expected.

 # The Origin of Neohybrids

Neohybrid hazels are plants created by making genetic crosses among several species of hazel. The crosses are made with the specific purpose of increasing useful genetic variations in the offspring, which is indeed happening at a remarkable rate—faster than expected. Please note, immediately, that neohybrids are in no way genetically modified organisms (GMOs). All the genes in neohybrids are *hazel* genes, rearranged into new patterns, but only via nature, not needles or gene guns. Neohybrids are instead the result of intentional use of an entirely natural phenomenon known as a hybrid swarm. It's something that happens often, in many species of plants and animals, with no humans involved. (There's more detail in chapter 12.)

When I first started writing and speaking about my work with hazels,

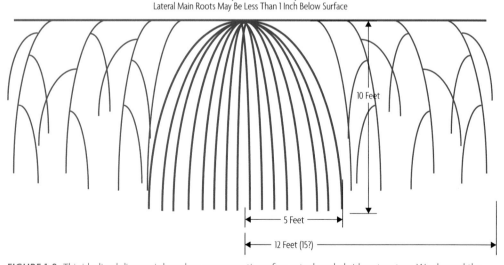

10-Year-Old Hybrid Hazel Root System
Lateral Main Roots May Be Less Than 1 Inch Below Surface

10 Feet

5 Feet

12 Feet (15?)

FIGURE 1-8. This idealized diagram is based on our excavation of an actual neohybrid root system. We showed the excavation to many professional plant scientists, who uniformly reacted with some (usually profane) version of "Holy cow!" The root system is much deeper, wider, and more extensive than anyone guessed.

FIGURE 1-9. Catkins in late summer, winter, and spring.

FIGURE 1-10. Compared with the normal receptive female flower (left), a neohybrid female flower has a very high number of stigmas (right). Our record cluster number so far is over 30; this multiplication of stigmas is likely part of the reason.

I used the term *hybrid hazels*. After several years, though, it became clear that this was confusing not only to farmers, but also to professional geneticists and other scientists who were not familiar with the additional layers of complexity, and the selection potentials, of crosses among species of plants when carried out through multiple generations. This type of breeding work hadn't been done this way before on purpose. So I coined the term *neohybrid*.

Traditional hybrid hazelnuts have been available to growers for at least 90

FIGURE 1-11. Denise illustrated a series showing the hazels in every season. Each detail is precisely true to life, and many of the details are painted using a brush with a single hair, under magnification. We're immensely proud of these; they are drawn from Badgersett neohybrids. Illustrations courtesy of Denise Walser-Kolar.

FIGURE 1-12. Food. Neohybrid hazel kernels.

years, probably more. In the nursery trade, these hybrids are still called hazelberts (a hazelnut-filbert cross), filazels (filbert-hazelnut cross; the first one is the female parent), and trazels (Turkish tree hazel–hazelnut cross). Various authorities will tell you the names have specific definitions. The problem for a geneticist is that the authorities usually disagree. The problem for the hazel grower is that nurseries are not at all careful about the use of these words. In practice the words have become meaningless. These older hybrid hazelnuts are different creatures from the neohybrids. They are typically crosses between only two species, and no one intends to cross them further; to me they are dead ends. Neohybrids are intended to be crosses among at least three species, and the process of recrossing and reselecting is not intended to end. The neohybrid gene pool, with all the diversity from three species available, will eventually be resolved into something new and quite different from wild and half-wild hazels—a new symbiont for humans. (Again, you can refer to chapter 12 for details.)

Our Three Species of Hazelnuts

Neohybrid hazels contain genes from up to three species of the genus *Corylus*. Understanding characteristics of the source species will help give a better idea of what characteristics may appear in hybrids among these species. Of course, creating neohybrids can bring out unexpected variations, and that's intentional—some of the new variants are useful.

AMERICAN HAZEL (*CORYLUS AMERICANA*) grows as a wild bush with a spreading crown throughout most of its range in the eastern half of

North America. Farther south they tend to be found in more mountainous areas; farther west they become riparian or are found in savannas, growing in prairie sod; but through much of the range they can be found anywhere there is woods, as long as it hasn't been grazed. Grazing a woods moderately for a couple of decades can knock the hazels out for decades or centuries, depending on the treatment of surrounding forests. Additionally, many pastures and farms had hazels forcibly removed (at great personal effort) in order to allow more room for crops or grazing. It has been a frequent comment from old farmers: "Hazels! I spent years digging them out! Now you want us to plant them?" Usually said with a wry grin.

Husks tend to cover the whole nut, often tightly. Stems don't get too thick, rarely exceeding 1 inch in diameter. Catkins point down in winter, and droop when shedding pollen. Nuts are usually roundish.

American hazels tend to kill grass growing beneath them. (Most neohybrids do likewise.) Evidence so far indicates to us that this is likely

FIGURE 1-13. Range of the American hazelnut (*C. americana*).

an effect both of the dense shade they create and of the chemical warfare plants engage in with one another known as allelopathy (al-el-OP-a-thy).

American hazel makes up the majority of genotypes and phenotypes in the current neohybrid gene pool.

BEAKED HAZEL (*CORYLUS CORNUTA*) extends farther north and much farther west than American. The bush appearance can be quite similar to American hazel, though in the forest understory *C. cornuta* may get taller. The leaves are more elongate. Husks have a single, tubular valve that tightly covers the nut and forms a long "beak." The husks have small, very sharp hairs on them, rather than the glandular or velvety hairs found on American and European hazels. Use leather gloves if you're picking or husking them! (So far neohybrids have not reverted to the beaked form.) Nuts are often more elongate and domed than American. The hilum is flat, a characteristic passed on to hybrids at a high rate and persisting in later generations, allowing us at times to identify parentage of new plants at a glance. That can be useful for

FIGURE 1-14. Range of the beaked hazelnut (*C. cornuta*).

suggesting what other characteristics we might look for. The catkins distinguish beaked from American; beaked catkins are no more than ½ inch long, upright, and sessile, in the nodes in winter, barely drooping when elongated.

None of the beaked hazels we've collected in Canada has thrived in Minnesota. After 20-plus years, most of them are no more than 3 feet tall, and rarely fruit. We attribute this to moving them almost 10 degrees of latitude, to a new spot where day-length patterns are quite different.

EUROPEAN HAZELNUT (*CORYLUS AVELLANA*) is a semi-domesticated species that forms the basis of the existing world hazelnut industry. It naturally grows as a bush, but is taller and with substantially stouter stems than either American or beaked hazel. In the bush form, wood may reach 3 to 4 inches in diameter; when forced into a tree, very much more.

Modern hazelnut culture often chooses to force the plant into a tree form, which allows easier collection of the nuts from the ground by machine once they fall free from the husk. As a semi-domesticated species, European hazelnuts tend to be bigger and develop full flavor more rapidly than the wild North American species, which may have little flavor at harvest.

Distribution descriptions for European hazelnut agree: They're found all over Europe and the Middle East. There is substantial variation in bush, nut, and clustering characteristics within the species; little of this wild variation has been used in modern commercial hazelnut cultivars, which rely heavily on Mediterranean hazels. Very small regions across the range, from Portugal to Kazakhstan at least, often grow local varieties that are well known in the immediate neighborhood. Always considered "the best in the world!" in each locale, they are typically unknown to anyone outside the tiny range. Not surprisingly, many are excellent and deserve much wider attention and breeding.

 ## Measured Neohybrid Potential Is Tremendous

The neohybrid hazelnuts now being developed are (mostly) small bushes. Compared with standard commercial hazels, they are far more amenable to machine harvesting, far less work to maintain, and intrinsically more productive.

Not only is the basic productivity of our best neohybrids already greater than traditional hazel industries report, but additional crop possibilities continue to appear: new flavors not known in any wild or domestic hazel, for example, and bushes that bear their nuts not in clusters of 5 or 6, but in clusters of more than 20 (and without the nuts becoming deformed).

Both fortunately and unfortunately, these new neohybrid bush hazels are significantly different from the hazels currently used in world hazel production.

Hazelnut Husks

American hazels have fleshy husks that totally enclose the nut during summer, and open gradually when ripe and drying, with multiple chemical weapons in both flesh and dense glandular hairs, for insects and for vertebrates.

Beaked hazel husks have a fleshy base, but seem to primarily use very small spines like cactus "hairs," with fewer chemical defenses.

European hazels typically have husks that are fleshy, not spiny, and do not completely cover the nut.

In the mixed gene pool, the variations recovered are huge, ranging from large fleshy husks with nuts nearly lost in them; to nuts with virtually no husk—and no defenses—at all; to hybrids with long, lacy ornamental husks.

Why all this husk stuff here? Because the nature of the husk fluctuates considerably in the neohybrids, and growers will need to make a choice as to which one they want to select for. What we can choose stretches much farther than the original species do; and what we choose will determine how well the nuts protect themselves from insects known and unknown, birds, mice, deer, and so on. White-tailed deer, for example, seem to dislike the taste of the green American-type husk; they very rarely eat them. They do eat the European types, those with much less husk to cope with. Birds prefer the huskless ones, too, sometimes attacking them well before they are ripe.

FIGURE 1-16. Wild beaked hazels in husks.

FIGURE 1-17. A hybrid with typical European husks.

FIGURE 1-18. A hybrid with nearly no husk.

FIGURE 1-15. Typical American hazel husks.

FIGURE 1-19. A hybrid ornamental husk.

The differences are fortunate because these plants are immensely more cold-hardy and disease-resistant. Their bush habit is an advantage too, because big trees require perpetual pruning and cannot ever be as productive as good bushes can. This holds true in all woody crops. Growers of everything from apples to peaches to squash and tomatoes now change to bush forms (sometimes called dwarfs) as quickly as breeders can provide them.

The unfortunate aspect of the differences is our lack of knowledge of how best to manage the neohybrids. Virtually all of the accumulated wisdom and information on "how to grow hazelnuts," based on maintaining standard tree cultivars, is proving near useless in managing plantings of these hybrids. The wild North American species added to the gene pool are simply very different in many ways.

While the crop now has some momentum and a committed core group of growers, it is nonetheless still in what can only be described as an embryonic state. Annual production from all growers is still under 10 tons, increasing only 1 to 3 tons per year, and a number of processes remain to be moved from the experimental state to the commercial. This is particularly true of post-harvest details.

 ## Where We Are Now

Today, there are many growers, universities, nonprofit organizations, and for-profit companies focused on building a hazelnut crop that is not based solely on European hazel genetics, and which can grow anywhere in North America. We've learned a great deal, and the concepts still look workable, as the University of Minnesota determined after examining the Badgersett Research Farm databases in 1998.[4] In fact, this review continues to be an important foundation of all research work on hazels in North America.

But we've also run into some unexpected developments, among them—not too surprisingly—new pests and problems. For instance, in 2002 our oldest hazels were about 20 years old. Now they're over 30 years old, and we have learned that 20-year-old trees are not actually "mature"; their growth form, growth rates, and fertility needs are different at age 30. This is not necessarily bad (we think the older plants need less fertilizer), but it means we're still adjusting the cultural protocols and parameters.

Here's another example of a surprise. Based on their parentage, we expected some breeding lines to reach a maximum height of 10 feet, but they're now 18 feet tall. Others are still the predicted 10 feet tall. We can now better predict what plant form and size will be when plants reach 20 years old, but unexpectedly we need to revise some specific breeding

FIGURE 1-20. This tour group is one of many Badgersett has hosted for dozens of professional organizations. Since 1995, we have welcomed individual scientists from China, India, the Philippines, Germany, Mexico, Spain, and more. Solid research connections grow from these beginnings.

schemes to reflect the new information. The biggest old plants get too big to be machined, too fast. We didn't know.

The neohybrids have been designed and selected to address many different problems, with built-in genetic answers. They are more productive (see figure 1-21 on page 19). The Badgersett hybrid hazels are far more widely adapted climatically than the European hazelnuts grown in the world industry, and incorporate disease resistance from a complex genetic base. Instead of cropping in alternate years, the selection program is finding genetics that allow the bushes to produce good crops every year.

Our search for useful genetic responses has been long, over 35 years now; intensive, with many thousands of data points compiled; and broad, as perhaps 30,000 individual neohybrids are now growing in Badgersett test plantings, out of approximately 140,000 planted and screened. (Of course, a couple thousand had their screening done mostly by gophers . . .)

The Lesson of Hybrid Corn: Change Is Difficult

I like to make an analogy between the adoption of neohybrid hazels and the transition from use of open-pollinated corn varieties to hybrid corn varieties. The transition to growing hybrid corn was not easy—in fact it could be described as painful. Farmers made many expensive mistakes

Breaking Records

Just as was done with the change from teosinte to maize, we are increasing the number of hazelnut seeds on one stem. Typical wild hazels usually have five or six nuts in a cluster. Among the neohybrids, we've identified multiple breeding lines that exhibit genetics for many more nuts per cluster, such as this cluster of 19 nuts. Our record is 32 (so far). Just increasing seed numbers does not automatically mean a bigger crop, of course; the plant must be adapted to the new fruiting. For example, a wild teosinte plant, with its tiny seed head, could never support or feed a whole modern corncob if you were able to genetically tack one on. The entire plant has to adapt: bigger roots, bigger stalk, more nutrients collected and translocated, and certainly other physiological adaptations we don't know about. They did it with maize; this says we can do similar things with hazels. Woody plant species in the tropics fruit in many physical ways not seen in wild temperate woody plants (cacao, for example, sets fruit right on the trunk and heavy branches); it's not any limitation in being "woody" that prevents hazels from bearing far more heavily. Or for bearing useful crops only four to five years after planting, instead of eight. Progress achieved so far indicates we are on the right track, with no biological barriers in front of us.

FIGURE 1-21. A cluster with 19 nuts, instead of 5. Our record is 32—the beginnings of a "hazel cob."

as they learned what hybrid corn was, what it would do, what it would not do, and what seed producers they could trust to provide quality seed. Hybrid corn still has detractors, but it is unquestionably more productive than open-pollinated strains (on average). And more profitable for the farmer. You can still find old farmers who remember the advent of hybrid corn, though they are getting fewer daily. Find some, if you can, and get them to tell you their stories—the useful education density is very high.

We're at an identical—and yet different—point in the development of hazelnuts as a modern agricultural staple. Identical in that once again, hybrid plants are being created, and sold to farmers, and new sellers of these new hybrids are appearing. What's different is that neohybrid

A Really Terrific Bull

Can you save your own seed from the neohybrid hazels? The answer is, "If you do it right." The caveats are that you must know what you're doing, and what your starting genetics are, and those genetics have to be worth carrying forward, or you can lose years. What's here in this book is a good start; then you'll need to get your hands on the plants and nuts, keep records, and learn to *see*. We'll help you.

That may all sound like a large burden for a farmer—but in fact, it's no more than dairy and beef farmers (and many others) do today regarding the genetics of their herds (and you're not required to do all this; yes, you can just grow them). Modern farmers are excellent and knowledgeable practical geneticists; they have to be. And as with any new endeavor, what

can seem incomprehensible when you start can become second nature with time, study, and practice. Proof: Recently my class in our short course demonstrated this beautifully. I'd put up a slide of a dairy bull's genetic chart, knowing that it looks entirely incomprehensible to non-dairy-farmers—yet every dairy owner I know reads them daily, for fun, in seconds. Unbeknownst to me, one of the folks in the class ran a dairy. While everyone else was boggling at the meaningless, endless abbreviations on the huge chart, he absently muttered "Wow! That's a really terrific bull . . . " Then, so all could hear: "Can I ask where you got that chart? I'd really like to check into that bull." He'd comprehended it all in about five seconds. You can get there with hazel genetics, too.

hazelnuts, as a genetic and biological phenomenon, are utterly not the same thing as hybrid corn. Hybrid corn is specifically highly uniform; each plant in the field is so similar genetically to the others they are nearly clones. Neohybrid seedlings, in contrast, are specifically highly diverse. In fact, neohybrid hazels are more diverse genetically than the wild species of hazels are.

The most common error farmers made with hybrid corn was attempting to save seed from their crop to plant in the next year. Saving seed from open-pollinated corn varieties works well, and was standard practice for millennia. But saving seeds from hybrid varieties doesn't work, because the genetics of the next hybrid generation change dramatically. It took a long time for many farmers to accept this; they tended to suspect that the seed companies just wanted to sell them new seed every year (way more details on genetics of all this in chapter 12).

An entirely different misconception stemming from hybrid corn was the astonishing idea that the seed couldn't be saved to produce a subsequent crop because "all hybrids are sterile." Like mules, people were thinking, I suppose. But hybrid corn is *not* sterile; it's just genetically erratic in the next generation.

A farmer who saved seed from a hybrid corn crop and planted it the following spring learned the hard lesson by watching the crop grown from the saved seed turn out to be a genetic jumble that yielded little to harvest. A painful mistake, but one that could be remedied the following year by buying new hybrid seed and planting again. Making the equivalent mistake with hybrid hazelnuts will be enormously more expensive for the grower. Because if you plant hazelnuts with poor genetics, it could take 10 years before you harvest your first crop—and discover that the plants perform poorly. In fact, hazels with bad genetics may never produce anything worth picking, let alone a crop.

It's a much more expensive mistake because the upfront cost of planting an acre of hybrid hazels is far, far higher than an acre of corn. So instead of losing a few hundred dollars, and one year's time and land use, a grower who plants hazels with bad genetics will have lost thousands of dollars and 10 or more years of production. Even more painful is the knowledge that, instead of having a hazelnut planting that will be productive for at least 50 years, the grower faces the grim reality of ripping out the unproductive bushes and replanting. None of us can afford this—not the growers, and not the new hazel crop.

This book is intended to give the reader a solid basic grounding in all the factors involved in commercial hazel production, from plant establishment and maintenance to marketing, and enough genetics to help you understand what might be possible in the next generation. My hope is that this book will help you avoid making mistakes like those suffered by corn farmers as they learned about hybrid corn. Change is not easy. And it can be hard to ignore your neighbor when he tells you "That's a crock! I've done stuff like that for years, and it just doesn't work that way!" These plants are—this crop is—*different*. Your neighbor may be entirely mistaken.

CHAPTER 2

The Mortal Sins
of Modern Agriculture

Most of the folks who decide to read this book are already familiar with the problems caused by modern agriculture. But not, of course, everyone. And often even those concerned about one grave difficulty may be unfamiliar with others. Now that you've been introduced to some of the inner working of neohybrid hazels, it's important to understand the consensus of experts on the status and challenges standard agriculture is struggling to cope with.

If you think "mortal sins" is over the top, you need to talk to an old farmer, or a young climate scientist. Agriculture today is the single most destructive thing our species does on our planet. We take all the best, most productive lands and plow them. Turn them to bare soil, usually twice a year, but sometimes four. . . . We are in mortal danger.

At the time I started focusing my attention on food systems, agronomy departments across the country were touting the endlessly rosy future of modern ag; flying tractors were just over the horizon. University crop scientists were pretty much wrapped up in their own progress, and uninterested in changing their directions, or thinking about new crops.

To my astonishment, in only 10 years, that attitude changed, dramatically. No longer muttering in corners, half the faculty were saying right out loud: "You know, we need to do this whole ag thing better. With less environmental damage." In a mere 20 years, the majority of the crop research world had changed from confident grins and "Yes, and we're already fixing that!" to "We need new ideas. A bunch of them. Really soon." Stated by people with very wrinkled brows. That's incredibly rapid change in academia, and evidence of the science and weight of the problems motivating it.

 ## One Part of the Solution

In 1988, I was invited to present a paper to the Second North American Conference on Preparing for Climate Change, in Washington, DC.[1] Struggling to describe what I was talking about, I came up with "woody agriculture." A formal definition: "the intensive production of agricultural staple commodities from highly domesticated woody perennial plants." We're talking about Food, to feed cities, from woody plants. The phrase has been adopted by others, and has come back to me from Germany, the UK, Australia, and even the United States.

Woody agriculture is not the same thing as agroforestry or permaculture. Agroforestry is traditionally concerned with producing food crops and trees for timber or similar uses at the same time—grazing beef cattle under black walnut trees, for instance. Permaculture, these days, can have different meanings for different folks; it's growing and evolving, in good ways. The word was suggested by J. Russell Smith's book *Tree Crops: A Permanent Agriculture*.[2] Beginning in 1978, Bill Mollison and Toby Hemenway spread the ideas[3] of permaculture widely around the world. Many books and websites by practitioners are easily available. Permaculture advocates a holistic, nature-driven farm, community, and life design, using trees, with many different crops growing with and under them. "Perennial polyculture," "food forests," and "forest gardens" are subsets.

Like everyone else working on simultaneous food and trees, I got my start by reading Smith's seminal book. Everyone from the early days of agroforestry (in all its permutations) and permaculture started with his insights. In his travels around the world, he repeatedly documented aboriginal peoples managing trees to feed their tribes—very successfully,

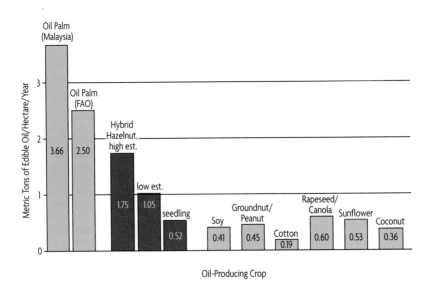

FIGURE 2-1. Metric tons of edible oil per hectare, world averages, around 2002. (It took me about two weeks of online research to put these statistics into comparable units at that time . . .) The hazel numbers are based on projections from research, not actual field production!

The Belt on Your Backpack

If getting our Food from trees is such a good idea—why didn't someone do it long ago? It's a fair question, despite the consistent cry of the Doubter—"If that would work, people would already be doing it!"—which history has shown repeatedly is just not so.

My own favorite proof of this is the backpack waist belt. Humans have been carrying loads on our backs since before we were humans (think mama chimp with baby on back). Through the ages, many cultures came up with ways to avoid having the load only on the shoulders—many shifting the load to the head and spine, as with the tumpline—but no one apparently ever successfully put the load on the hips. If you've ever backpacked, you know that no one in their right mind today would carry heavy loads without a padded waist belt. But humans did so for tens of thousands of years, without, so far as we know, the concept of the waist belt ever occurring to anyone. Ötzi, the Iceman from 5,000 years ago, was uncovered when global warming melted the ice on his mountaintop. He carried a set of tools showing great sophistication; a different species

of wood for each different need. His backpack was a single hazel rod bent into a U, with two larch crossties and no waist belt. The ancient Egyptians, ancient Greeks, ancient Chinese, and so on all spent a great deal of time thinking about engineering and tools, but it was only when Dick Kelty and his friend Clay Sherman came up with the belt idea while hiking in 1951, and Kelty worked for years to make it reality,[4] that our species started to use this obvious and incredibly useful idea. Obvious—after the fact. Like gravity is obvious—*after* Newton and his apple. (I bought my first external-frame padded-waist-belt backpack in 1967, and took it to Europe in the summer of 1968. Local hikers on mountain trails in Switzerland, Austria, and Germany had never seen one before. They tried to buy it off my back.)

If humans have overlooked something so necessary, so basically useful, and so *simple*—for *millennia*—what else have we been overlooking? That would make an excellent koan for meditation. And who knows what you might come up with?

and without destroying the landscape. Our own "modern" agriculture is much more destructive, as one of his chapters, "Corn, The Killer of Continents," points out.

Smith was a geographer, hybridizing, if you please, his field with both anthropology and economics. He was one of the co-founders of both the National Geographic Society and the Northern Nut Growers Association. But he was born too soon to be aware of the astonishing power of modern genetics and modern evolutionary theory.

Smith's book presented a possible answer to a question I had already formulated: *How can we design an agricultural ecosystem that allows us to skim enough energy to feed ourselves without ultimately collapsing the ecosystem?* Trees can be domesticated, using modern tools and our greatly increased understandings of genetics and evolution. No tree ever has been domesticated, in my opinion. At least not to the extent that maize or rice have been changed from wild forms to fit human requirements. Apples and grapes? Semi-domesticated, at best; they require tremendous amounts of labor as well as pruning and vine supports throughout their life spans. And most traditional tree crops do not produce what we're calling Food: reliable, non-perishable staples, in the quantities that can feed cities.

Neither olives nor coconuts are domesticated in our opinion. Since that first paper in 1988, oil palm has emerged as a fully industrialized tree crop; unfortunately, it uses maximum fossil fuel and pesticide inputs to establish huge monocultures where tropical forest used to grow. But oil palm does illustrate that trees can produce Food. They are genetically capable of long-term very high food production. And the most productive oil palms? Are species hybrids.

Hazels are the woody perennial seed (oil and protein) crop that is farthest along the track to becoming truly domesticated and capable of producing Food that fits our needs today. And it does so without repeating the problems of modern industrial agriculture. Let's look at how that works.

Soil Lost

The plow, in all its variations, has been the great weapon in agriculture's war on "weeds." Turn the soil over, and the native plants die so the crop can live. But with no plant cover during so much of the year, the loose soil is free to wash away when it rains or blow away when winds come up. Attempts to keep some cover on the ground, such as "low-till" and "no-till" practices, do better—but by killing the weeds with chemical herbicides.

Grain agriculture leads to most of the world's best soils being naked, unprotected from wind and water erosion, for approximately half of the year. Even in the parts of China where they raise two crops of rice and one crop of winter wheat every year, tillage, planting, and crop-drying stages (when no photosynthesis goes on) still have to take place.

The result is immense quantities of soil washing off agricultural fields. The result is cities such as Ostia, which was the main seaport of Rome around 400 BCE. It now lies almost 2 miles inland, due to silt carried by the Tiber River from plowing more hills to feed growing cities.[5]

There is now a large industry whose entire business is to measure how much soil we are losing due to agriculture, where we are losing it and how, and what the trends in soil loss are day by day, year by year, century by century. There are differences of opinion among these experts, but typically new measuring capabilities lead to statements like this one, from The Environmental Working Group and Iowa State University, in 2011:

> In some places in Iowa, recent storms have triggered soil losses that were 12 times greater than the federal government's average

FIGURE 2-2. This is a wide-angle photograph of a new Iowa cornfield. Nothing is alive in this field except the emerging corn. If anhydrous ammonia fertilizer was used (standard practice in the United States), most of the worms and other soil organisms have also been killed. This is the overwhelming reality of modern agriculture.

for the state, stripping up to 64 tons of soil per acre from the land, according to researchers using the new techniques. In contrast to the reassuring statewide averages, the researchers' data indicate that farmland in 440 Iowa townships encompassing more than 10 million acres eroded faster in 2007 than the "sustainable" rate. In 220 townships totaling 6 million acres, the rate of soil loss was twice the "sustainable" level.[6]

Restoring Soil

No soil blows away or washes off from a field of hazels. Hazel bushes grow 10 to 16 feet tall—a field of them cuts wind drastically. The ground beneath them is held tightly by roots, 365 days a year, both by hazel roots and by the roots of many grasses, herbs, and forbs that grow between the trees. We've even found jack-in-the-pulpits (*Arisaema triphyllum*) establishing themselves by the dozens, planted by resident birds, under

FIGURE 2-3. Yes, those are our hazel stems in the background.

our 25-year-old hazels. This wildflower is usually considered an indicator of rich, undisturbed woodlands.

When we excavated a hazel root system in order to measure it, we found that the exposed roots held the soil so tightly that we had to resort to using a 90 PSI water jet to wash the soil off the roots.

A planting of hazels will capture blowing soil. When the snow-drifts melt slowly under our hazels we can see the black lines of soil carried there by the wind from someone else's plowed field. In spring thaws and even flooding rains, water sinks into the soil—which is riddled with undisturbed earthworm and soil insect burrows—and runs freely down the thousands of bathtub drains kindly installed under the hazels by mice, shrews, and moles. All of this benefits the hazel crop, in dozens of ways.

Water Contaminated

Remember the naked soil? Conventional farmers apply fertilizer—enough to support a full year's growth for a bumper crop of corn—to soil *before* planting the crop, when there are no roots to absorb it or hold it in place. When it rains, some of that fertilizer dissolves in the rainwater and is leached away. That loss is part of the annual fertilizer calculation: Farmers factor in an amount they expect to lose, add that to the amount the plants actually need, and that's what they apply. Excess fertilizer winds up in rivers and oceans, but also in our groundwater. So do herbicides and pesticides.[7] Excess nitrogen from the Corn Belt is now acknowledged to be the primary cause of the "Dead Zone" in the Gulf of Mexico.[8]

There is also the very large problem of irrigation water for agriculture. According to the United Nations International Fund for Agricultural Development, 70 percent of the world's fresh water is used for irrigation.[9] Much of that will be used by the plants, but much also will simply evaporate in storage, transport, or from bare soil. Much will also run off into streams, or percolate down into aquifers. All over the world, water tables and aquifers are dropping and wells are going dry as cities and agriculture use more and more subsurface water. The drops in aquifers around the world are horrifying; one of the best studied is the one on which Beijing relies. "The water table under the capital has dropped by 300 meters (nearly 1,000 feet) since the 1970s."[10] I find that terrifying.

Naked, tractor-tilled croplands cannot absorb heavy rains because the subsoils become packed more and more densely until they turn into a hardpan layer, which even the plow cannot break up. Rains no longer infiltrate to recharge aquifers as they used to, but run off into surface drainages. All of that water may carry fertilizer and/or pesticides. Irrigation can also cause deserts when the evaporating water carries salts up to the soil surface, eventually making crop growth impossible.[11] The process is known as salinization, and land typically continues to be irrigated, with declining crop yields, until it is actually abandoned.

Clean and Abundant Water

We have done extensive measurement of the results of fertilizing neohybrid hazels. A grant from the Minnesota Department of Agriculture's Energy and Sustainable Agriculture Department allowed us to do multiple controlled experiments. I can't describe all the details here (it might be a book in itself), but I do want to describe one aspect of one experiment. Working in consultation with Dr. Bert Swanson of the University of Minnesota Department of Horticultural Science, we divided six different 300-foot rows of hazels into four treatment groups. The control group received no fertilizer; the other three received the equivalents of 100, 200, or 300 pounds of nitrogen per acre. (For comparison, most corn farmers in our locality apply between 100 and 200 pounds of nitrogen per acre each year, with a few using as much as 300 pounds, to really push their yields.) Soil tests before the application indicated about 5 parts per million (ppm) of nitrogen in the top 6 inches of soil, with 0 ppm at 12 and 18 inches. The fertilizer was applied on September 1, a date usually considered too late, even risky, for most woody plants. It often happens that late fertilization of woody species can create a new flush of growth, which may freeze badly even in normal winters. Our past experience gave us confidence that the late application would not harm the hazels. We were fortunate that the winter turned out to be a "test" winter, one of extreme cold for our region (−42°F, or −41°C). There were no effects on the cold hardiness of the hazels, and no symptoms resembling over-fertilization appeared in the 1994 growing season. In late August 1994, 11 months after the initial fertilizer application, in the rows treated with the "extremely" high rate of 300 pounds of nitrogen per acre, the soil nitrogen measurements were: at 6 inches deep, about 15 ppm; at 12 inches deep, 5 ppm; and at 18 inches deep 0 ppm. To interpret the numbers: There were still "good" nitrogen levels in the top 6 inches, but nitrogen was very low at 12 inches and nonexistent at 18 inches. The hazels had allowed no fertilizer to escape into groundwater after 11 months of snowmelt and rainfall. That's not very surprising, given that the hazels have very large and permanent root systems, always in place, and capable of active nutrient uptake even in winter, below the frost line.

As for pesticides, we don't use *any,* so none can escape into the water. And here, in a climate with approximately 28 to 30 inches of precipitation a year, we have never irrigated the hazels, except during the establishment year.

As for abundance of water, untilled fields of hazel bushes develop soils that are alive. They support an abundance of symbiotic species: fungi, bacteria, beetles, worms, millipedes—all of which leave tunnels of various sizes behind them, which allow water to soak deeper, faster. On a much larger scale, mice, voles, shrews, moles, and ground squirrels create and maintain what amounts to a natural flood drainage system. When water pools on the surface, even when the ground is frozen, that water will eventually find its way into a rodent-made drain and run down it like a bathtub drain. The rodents are no dummies: Their deep dens are always below the frost, and they have several different ways to prevent the nests from ever flooding. Incoming waters run down all these holes, reach unfrozen soils, and soak in.

Air Polluted

The word *pollution* is used to refer to a multitude of very different, even unrelated sins. Anything we put into the air that doesn't belong there, that causes harm, is pollution. Modern agriculture's effects on air are complex, but when we carefully add them up the results are very negative. There's the straightforward air pollution that results from application of pesticides. If you've ever driven past a field where something was being sprayed, and the wind was blowing the wrong way—you've smelled the fact that not all of it goes on the ground; some of it stays in the air.

Modern agriculture is one of the largest contributors of excess greenhouse gases (both CO_2 and methane) to the atmosphere. Advocates of big agriculture will point out that crops capture CO_2 from the atmosphere and release oxygen, both of which are beneficial. What they fail to state is that the original ecosystem displaced by the row crop was capable of far more photosynthesis over the year than the monoculture planting, which is actively photosynthesizing for only half the year.

In addition, when soils are turned over—tilled—buried soil organic material is brought to the surface and oxidized, turning to CO_2 and adding to the burden of greenhouse gases (besides decreasing the fertility of the soil, and its ability to hold nutrients). Then there's the CO_2 generated by burning of fossil fuels to operate the tractors, sprayers, and combines, plus nitrous oxides from diesel motors. All these contributions are a very large part of greenhouse gases from human activity.[12]

Still another negative contribution from pesticide application is the lost potential for photosynthesis. Pesticide data have become more elusive because in 2001, the US Environmental Protection Agency stopped reporting how much is used annually. The number is somewhere around 2.5 million tons (using American measure—1 ton equals 2,000 pounds).[13] That would be 5 billion pounds of toxins, much of which is designed to kill plants, which again reduces the potential photosynthetic contributions of the acreages that have been sprayed. It gets very hard to untangle, but it's all very, very bad for your grandchildren. Farmers know this. But most of them are truly trapped in a system that has no real exit path.

Air the Planet Can Breathe

Hard scientific measurements and comparisons of one crop system with another are difficult and slippery, and have been known to result in heated academic feuds: "You didn't include X." "You measured Y incorrectly." I prefer to use one reasonably measurable indicator of efficiency: how much new biomass (here used in the ecologist's sense of "everything that is or was alive") does this crop produce in a year? Crops are designed to capture sunlight. Some of the sunlight they capture is used by the plant itself during the year, for metabolism; some remains at the end of the year as harvested food crop, or dead leaves, wood, roots, or stalks. The total of everything left at the end of the year is the "annual biomass production." Including roots, leaves, stems, and all, trees can make about three times as much biomass in one year as a maize crop can.[14] Which means—approximately—that the woody plant took three times as much carbon, as CO_2, out of the air. There are many ramifications to this, one of which is that if we change enough plowed crop lands into woody crops, we could actually start to pull enough carbon out of the atmosphere to make a real difference in greenhouse gases. The *ifs* and *maybes* are huge. But if enough woody crops were adopted, we could see annual greenhouse gas numbers start to come down instead of steadily increase.

In part, this depends on whether a given crop is a C_4 plant (these plants can capture four carbon atoms per photosynthetic "cycle") or a C_3 plant, which captures only three such atoms. Maize, sugarcane, sorghum, millet, and amaranth are C_4 food crops; all other crops are C_3. Currently there is great academic enthusiasm for altering our food crops so they can use the C_4 pathway instead of C_3. One of the potential fallacies of the C_4-to-C_3 comparisons, however, is that the C_4 advocates are assuming an equal amount of leaf area (in other words, chlorophyll) for the C_4 and the C_3 crops. But if the comparison is between a C_4 crop and a *woody* crop, that is an inappropriate assumption. Compare leaf area in neohybrid hazels to corn. The mature height of the hazels is around twice that of corn. Hazels are also fully leafed out and photosynthetic for a much greater portion of the year. When sunlight is at its most intense at the summer solstice, it is common in the Corn Belt for sunlight to be shining on bare soil in cornfields. In a hazel planting, the trees are fully leafed and photosynthesizing at maximal capacity a month before that. And in the fall, a month or more after this year's maize is dead and dry, hazels are still photosynthesizing, pulling CO_2 out of the air. (For more about the C_4–C_3 comparisons, see chapter 13.)

At the outset of Badgersett hazel research over 35 years ago, I didn't spray for pests because I did not know what pests existed, or how serious they were. What I've found by not spraying is that (so far!) every pest that has showed up, has increased, and then has decreased to marginal, usually non-economic levels of crop damage. This is a subject that will recur, with more detail, throughout this book. It is in fact one of the true paradigm shifts in this crop—we do not "control" pests, we manage them, and without toxins. No pesticides—and very likely, not ever.

And remember, sprayed poisons always destroy some non-target species also. Even "bio-pesticides" usually presented to the public as non-toxic and specific, such as *Bacillus thuringiensis kurstaki*, actually kill not just the target but also many butterfly and moth species—and can harm connected species that feed on them.[15] Leading to—

Biodiversity Destroyed

Biodiversity refers to the number of species in an ecosystem at its most wide reaching, including the bacteria, fungi, and even viruses. This web of interconnections is absolutely real; every organism in the system depends on several more, is depended on by others, and interacts with many, many more. There is broad agreement among scientists that maintaining biodiversity in our future is possibly as important to our own survival as any other issue, including climate change.[16] We will not survive if we cause the collapse of too many ecosystems around the planet.[17]

Modern agriculture, unfortunately including such tree crops as olives, oil palms, apples, walnuts, and the present hazelnut industry, creates artificial biological deserts; nothing is allowed to live except the crop. The basis of modern agriculture is total warfare on all species except the crop. The argument is put forth that annual tillage and use of chemicals is the only way to be competitive in the market.

In the hazelnut industry in Oregon, it has been standard practice to spray copper sulfate to kill moss and lichens growing on branches. The moss and lichens don't harm the trees, but it's argued that in a snow or ice storm branches with more moss and lichen are more likely to break off, taking the fruit buds that would produce the following season's crop with them.[18]

Hundreds of different organisms live in that moss and lichen: insects, beneficial and otherwise, bacteria, fungi, isopods—the list is mind boggling. And how do they all interact, naturally? We don't know; the complexity is beyond our ability to sensibly study it. When we spray to kill the moss and lichens, what else are we destroying? We don't know.

That is the hard reality of modern agriculture: We do not even know what we are losing. To put it on a purely human benefits basis, how many new cancer drugs, malaria cures, new miracle diet fruits, and so forth have we destroyed without even realizing it when some rare plant, or millipede, or fungus was pushed into extinction without ever being recognized as vast stretches of Amazon jungle and Argentine grasslands were cleared and plowed to raise grass for beef and soybeans? We'll never know, but consider that one of the most powerful drugs in our anti-cancer arsenal, Taxol, was discovered in the bark of an extremely slow-growing, near-threatened tree, the Pacific yew. (Taxol still cannot be synthesized economically; a way was found to use European yew to produce a precursor molecule.)

This is not only a matter of human utility or economics; it is also an ethical and spiritual problem. And keep in mind, modern agriculture is no longer only about food; increasingly land is dedicated to or cleared for fuel crops. Can it be right for us to destroy entire species? I find the concept horrifying. Not just because the damage to the world, past and potential, is horrifying, but also because I, too, am a member of another species.

Biodiversity and Agriculture Integrated

Since we do not spray them, and do not till after establishment, hazels provide habitat for everything. The soil is undisturbed and unpoisoned,

A Very Special Spider

The best example I can give you of the extent of the diversity in the hazels is the spider in figure 2-4.

What spider? I hear you say. Look very closely. See that bud at the top of the dead hazel twig? In 2004, my son Perry was helping collect data on the hazels during harvest. He is a spider enthusiast. He noticed a scrap of fine web, but he couldn't find the spider. He called me over, we both looked, and we couldn't find it. Eventually one of us spotted it, and I was immediately ecstatic. It was perfectly camouflaged—the sign of long co-evolution with its habitat. It is clearly a dead hazel bud mimic. It's about 85 percent probable that the only place this species exists is on hazel bushes.

As a student of animal behavior I had studied mimicry across many taxa, from birds, to orchids, to insects. This was spectacular. No one yet has been able to identify this spider for us; quite possibly it's a new species. How many spider scientists go prospecting for spiders on the tips of hazel twigs? I came back the next morning, and the spider was still on the same twig, but at the tip. I spent another hour observing and photographing. The final proof that this tiny predator is a bud mimic came when Perry and I started to nudge it a little, to see how it would move. It didn't want to leave that twig tip. Poked gently with a thin grass stem, it would lift up its body halfway, shift position very slightly, and sit right back down. This happened several times. When we continued to nudge, eventually it scampered down the stick—to the next bud scar, where it once again sat, compacted its legs, and became a bud. The only place it would settle was on a bud node; it knew where it was, and it knew it was hidden. After it repeated the next-bud-down-and-stop routine three or four times, it ran back up to its original position at the top. We decided that was enough poking for knowledge, and left the spider there, one leg feeling the tension on its web.

It was gone the next day, and we've never spotted one again. I'll bet they're there, though, just hidden in their own cloak of invisibility.

FIGURE 2-4.

FIGURE 2-5.

FIGURE 2-6.

the above-ground plants survive for years. In a mature field setting, wood from the bushes is harvested only on a rotating basis, so that mature bushes are always available for species to survive on, or move to.

Birds nest, and feed, and hunt, and flock in migration, in the hazels; all year. We see new species every year. Tree frogs live in the hazels; eating bugs for us.

The entire world holds intricately adapted biodiversity in our natural ecosystems, and we do not even know it is there. Every species is a part of the whole, and every species precious. The urgency of preserving as much biodiversity as possible may be the aspect of climate change that has been most difficult for scientists to convey to the public. It's easy to interest the world in saving the pandas. It has been painfully difficult to get the world to see that invisible web. But without our web, we will fall.

Starvation in Drought

Drought. Historically, this word is associated with another—famine. This ancient periodic curse of our species is far from extinguished; the miracle of modern agriculture has not banished it. Seeds that must be planted in bare soil every year have a series of requirements that must be met for the crop to succeed. First the soil must become dry enough to till. Once the ground has been tilled and planted, the seed must receive adequate rain (or irrigation) to germinate the seed and support plant development while roots are small. As the roots grow deeper, they may be able to tap soil moisture and do without rainfall for some time—if there is any moisture deep in the soil. And at the end of the season, there must be dry weather so seed can dry to storability, and harvesters can gather it.

If at any stage the weather is outside the crop's requirements, there is danger of crop failure. During a drought, the seed may lie ungerminated. Even a short drought can cause trouble. If the seed germinates later than usual, the yield will decrease—or it may germinate late enough that the crop no longer has time to mature at all. Or the crop may start well, but dry conditions in the middle of the season may cause it to die only half grown.

Survival in Drought

I've seen all those things happen to my neighbors' corn over the years. In 1988, the USDA classified our area as in extreme drought. My neighbors' corn grew about 3 feet tall—and died. Our unirrigated hazels bore a near-normal crop; when I analyzed the harvest data, they showed that the average whole nut weight was down by around 10 percent, but most of that came from a decrease in shell, not kernel. The ground became dust 2 feet down. The hazels have roots 10 feet deep, and were still reaching adequate moisture.

The next year, 1989, we still had extreme drought, but with a slightly different pattern. Instead of no rain ever, we began to have ¼-inch rainfalls once a week. Just barely enough so that some corn germinated. The top 2 inches of soil was moist, and a small corn crop was grown. I expected nothing of the hazels, given the stress they'd undergone the previous year, but they still produced a crop. This was when I started to pay serious attention to their potential.

At every stage of the growing year, the requirements and tolerances necessary for successful annual crops are much narrower than the tolerances for deep-rooted woody plants. For the hazels there are no germination requirements; wet or dry is of no significance. Moderate annual droughts that can kill or damage young row crops are not even noticed by the hazels. Much more serious regional drought events are survived and tolerated, and the hazels may even produce crops.

Starvation in Flood

The story is the same with flooding, though inverted. In very wet springs, planting can become impossible. If newly emerged corn seedlings are hit with enough rain to submerge the leaves for a day, the plants will die. Much more serious flooding sometimes occurs: several feet of water running over the field, or water standing on crop fields several feet deep for weeks. Such conditions also mean the end of that year's row crops.

Harvesting row crops can be impossible when the ground is too wet. Machinery can bog, and wet soils can compact, leaving long-lasting damage. In these conditions, the crops simply remain unharvested, food lost.

Wet conditions have an impact on drying and storage of conventional commodity crops, too. Grain crops and pulses need to be dried fairly quickly and kept quite dry until use. It doesn't happen in the developed world much now, but in less developed countries fuel for drying crops is not affordable, and in years when sunshine doesn't happen on time, crops will spoil. Likewise, due to inadequate storage facilities, in a rainy year stored dry grains may get wet in storage. If it's a time of year when solar drying is no longer possible, crops are lost.

Tolerant of Flood

For hazels, the kind of climate events that destroy modern row crops are of no consequence. A few inches of standing water has no effect on a 1- to 12-foot-tall hazel bush. We know by long testing that most neohybrids tolerate chronically wet soils, including occasional flood. We have some hazels planted in a pond bottom that floods every spring, sometimes long enough to attract frogs trying to breed. It has also held 2 to 3 feet of water during major local flooding (we've been an official federal disaster area twice in the past five years for flooding) with no visible effect on the hazels, bush or crop. Reports from growers state that if mature hazels are under 2 to 3 feet of water for a *month* or three—yes, the submerged leaves and nuts drop off. But the leaves and nuts above the water survive just fine, as do the plants.

We plant the ground between neohybrid hazel bushes in some kind of perennial grass or legume cover. This cover, underlain with extensive hazel root networks, allows harvest to go on until the flooding becomes truly extreme.

And when it comes to survival in storage, neohybrid hazels are remarkable. Wild hazels, which make up much of our genetics, can be buried by a mouse or squirrel, stay underground for as long as three years, surrounded by soil—undergoing multiple occurrences of freezing, thawing, soaking up rain water, and drying—and still be sound and capable of germination. They will survive spoilage conditions that would quickly turn any grain or bean into black slime. If they should get wet in storage, they can be redried. And wet and dried again. In many commercial hazels this extreme resistance to spoilage has been mostly lost; it is not a characteristic that has been selected for in general. We pay attention to it, and many neohybrids do retain the resistance to spoilage of their wild progenitors.

Too Good to Be True

Yes, we know: It all sounds like too much. Too good. Has to be snake oil, right?

We've certainly heard that, many times, but we would ask you to just keep reading; digest this new information for a while. See if you can get to the end of the book and let it simmer in the back of your mind. And keep these points in mind as you do:

- It took me around 15 years to begin to believe hazels might work as a regional crop.
- It took 25 years for me to think the word *believe*—which is taboo among scientists—and 30 years to say it out loud.
- Not *one* of my very closest friends—most of whom are good-to-exceptional scientists—thought there could be anything really to all this until they *saw* the neohybrid hazels with their own eyes. But then they all saw real potential, and became supporters. Yes; it's all hard to believe.
- This *is* a true *paradigm* difference. That term has been bandied and abused until its actual meaning has become vague for most of us. It means "world-changingly different." Like the difference between sailing ships and steamships. Horse-powered transport and automobiles or trains. Surface mail and telegraph, telephone, radio. Every one of those changes was faced with enormous skepticism and took decades for people to absorb. This would be as big a change. Hard to believe. But a true paradigm change is what we need right now. Which can, of course, make it more difficult to accept. Give it some time. *Check* everything. Maybe try it a little and see where it goes.

CHAPTER 3

The World Hazelnut Industry

The overwhelming majority of the world hazelnut industry does not meet the standards of sustainable agriculture. Production methods are based on clonal monoculture, minimum genetic diversity, and high inputs for tillage, machinery, and pesticides. Production is expanding and new regions are coming into production. However, the industry overall is very conservative about markets. The traditional hazelnut markets around the world are luxury consumption, often related to holidays and gifts. *No one today views hazels as a significant food crop.*

It is easy to find data on the world hazelnut industry. Be wary, though, about accuracy. First—check the date of what you're looking at. If you can't find a date on the document, don't waste your time; it could easily be 10 years old, even 20. The market has changed since then. One good source of information is the UN Food and Agriculture Organization (FAO), which keeps databases on everything you can think of. Their accuracy is the best they can manage (far from perfect, since some governments literally make up their data). Find FAOSTAT on the Internet (http://faostat3.fao.org) and dive in; it's highly educational.

Second—be aware that the world hazelnut market is profitable, but established uses for the nuts are quite restricted compared with agricultural commodities such as maize and soybeans, which are used for industrial purposes—from ink to animal feed to food additives like lecithin or cornstarch—far more than as food directly for humans. This means that hazel producers are in fierce competition with one another for small, established markets. The traditional hazel-producing regions have highly professional full-time marketing programs—and they all spin all the information in their own favor. Oregon, Turkey, and Italy all claim that they produce the hazels that set all world standards for excellence.

Neohybrid hazels in the Midwest have not made it into any official reports yet, because there is no industrial-scale production yet—and because Badgersett is very serious about our science. We don't *like* spin, professionally or personally.

In that spirit, keep firmly in mind that neohybrid hazelnuts are generally unlike the world crop. That should surprise no one; the genetics are

The Alternate Bearing Effect

The majority of hazel cultivars currently used in commercial production are susceptible to alternate-year bearing—the tendency to have a small crop the year after a large one. Breeders of traditional hazels all work at decreasing that tendency, but it is still a significant factor; production in Turkey and Oregon has sometimes dropped as much as 50 percent the year after a heavy crop. This can cause price swings, because typically little crop is carried over from the previous year. This is one of the reasons the various regions' statistical percentage of world production can vary from report to report.

radically different. Some of the neohybrids do strongly resemble those from other regions, but at this point with seedling-based plantings, the majority of neohybrids have smaller kernels and quite variable flavor.

During my years as an officer and board member of the Northern Nut Growers Association, I met official representatives from many hazelnut regions around the world, both growers and scientists. I acquired a good deal of information that does not show up in official reports. I've included some of that "not in the official reports" information here, as well as my best guesses as to what the numbers really are.

Turkey

Turkey is by far the largest hazelnut producer in the world. I usually give their percentage of production as around 60 percent, although Turkey's report for 2013 claims 75 percent. One of the confusions in the data is that some report "production" and others "exports"—and they often do not tell you which. Turkey's share of international trade has been around 80 percent for a good while. The country's impact on prices is therefore large. A factor adding to the volatility of prices is that sometimes Turkish exports can vary for political reasons, not weather or natural fluctuations.

It is often assumed that Turkish production is based on a species of hazel known as Turkish hazel (*Corylus colurna*). This is not the case. All commercial production in Turkey is from *Corylus avellana*, some of European origin and some from *C. avellana* native to Turkey. Most production occurs in the regions where climate is tempered by the Black Sea. Plantings are increasingly intensive and mechanized, but a great deal of culture is still traditional bush-form plants, and harvest is by hand. The government encourages expansion of plantings and supports excellent research. A reasonable long-term average for production is approximately 400,000 tons per year; a recent high reached 660,000. For a comparison, US soybean production in 2013 was around 3 billion bushels, equivalent to 90,000,000 tons. In terms of food production, hazelnuts are insignificant today; they are specifically a luxury, not a staple. Yet. See chapter 14 for data on soybean production in 1919, when it was a new crop in this country.

 ## Italy

Italy is normally the number two producer, with around 10 percent of world production. Italy is also one of the larger consumers of hazelnuts, however, and it imports from Turkey to supply its large hazelnut-chocolate export industries: Nutella, Ferrero Rocher, and multiple others aimed at elite markets. Italy is also expanding plantings and supporting research; newer plantings are almost always clonal and mechanized, but hazelnut production in Italy is thousands of years old, and traditional practices survive, both as farmer and as boutique operations.

FIGURE 3-1. In mature traditional hazel culture, plants grow as multistemmed bushes, not single-stemmed trees. This growth form is not as machine-friendly or highly productive, since it tends to close out sun. Harvest is by hand, often by families that have been harvesting the same hazels for hundreds of years—sometimes possibly thousands. This traditional pattern can be found from Turkey across Europe and into the Balkans, Ukraine, and farther east. Photo from Joadl, Wikimedia.

United States

The United States is the world's number three producer, with an average of some 30,000 tons per year, varying from 20,000 to 40,000. Almost all of the commercial hazelnut growers are located in Oregon's Willamette Valley, although there is also some production in Washington. In the past few years, US producers have dramatically increased their exports, to the point that almost 50 percent of the crop is now sent overseas, most of it to China. That market is still expanding. Attempts to export to Germany and Italy have had very limited success; the quality of the nuts is different enough that chocolatiers are reluctant to use them.

The Oregon/Washington industry moved to limit grower production some years ago, so that established markets would not become saturated. Toward that end, growers sought a federal marketing order (number 982)[1] establishing a marketing board that regulates standards and production. This marketing order applies only to hazelnuts grown in Oregon and Washington, however. Only clonal varieties are planted. The hazels are shaped into single trunk trees, via pruning and spray; allowed to drop to the ground (which must be flat and smooth) at ripening; and mechanically swept up for harvest. The industry is expanding at the moment, with 2,000 to 3,000 new acres planted per year.

FIGURE 3-2. Not the biggest and most beautiful Oregon field, perhaps, but it's fairly typical in the flat field floor to allow sweeping, and the shaping of the hazels into trees, not bushes. Photo from M. O. Stevens, Wikimedia,.

Since the 1960s, an epidemic of eastern filbert blight (EFB) has been a problem for US producers. (For details about this disease, see chapter 6.) At the onset of the epidemic, all the genetics in use were pure European hazel—and up until very recently all of Southern European origin. This was not a problem when the Oregon industry was first established, because the fungus was not present. It was known that the germplasm wasn't resistant to the blight fungus, but all concerned believed it was unlikely to reach the western United States. Unfortunately, it did, and the susceptible trees began to die. Spraying fungicide every year forever is expensive and in many cases ineffective; as a result entire fields have had to be bulldozed. Oregon State University conducts the research for the industry and has worked hard at developing new cultivars, but it is an uphill struggle. The first "resistant" cultivars released proved inadequate in the field and have been largely abandoned; a second generation with additional resistance genes is now in release and being planted. The genetic diversity of the fungus in Oregon is nil; it has been determined that the entire fungus population is of a single genetic origin. In places where the fungus is native, however, it is highly genetically diverse,[2] with a much greater ability to adapt. This has proved a problem for the Oregon breeding program. So far, even the best Oregon State new releases have failed to survive in the Eastern US; all succumb to the fungus, usually rather quickly (my personal observation, and that of others). If and when the more genetically

One of Those Things Nobody Knows

As founding president of the American Chestnut Foundation, I'm very sensitive to the movement of disease fungi; and I can get very cranky when someone is being insufficiently careful about spread.

Badgersett is a licensed nursery, inspected and certified by the State of Minnesota. It is not allowed to ship hazels into Oregon, Washington, or California, because it's located in a region where eastern filbert blight is endemic. That quarantine was established in 1929 with the intention of keeping EFB out of those three states. Makes sense.

But. Some years ago, as I was checking the annual changes in nursery regulations, I discovered that the quarantine had been lifted. I was now allowed to ship Badgersett's certainly EFB-infected material anywhere.

I don't know what the regulators were thinking, but I knew, professionally, that it was an enormous mistake. It was well known at that time that there was virtually no genetic diversity in the EFB fungi present in Oregon and Washington, but that the fungi in the eastern United States were highly diverse. Any new transport of the fungus could very easily make the existing fungus in the western states far more pathogenic.

I called a friend who lives in Oregon and writes about hazel production there, explained the mistake, and raised a little hell. He quickly called the Oregon regulators, and in a week the quarantine was reinstated. It was the right thing to do.

diverse fungus reaches the Pacific Northwest, the bulldozing will have to be repeated. We hope the quarantine holds, but we know from many other events that private individuals sometimes transfer plants illicitly, unaware or unbelieving.

 ## Other Production Areas

Rankings have changed a good deal in the past decade. Currently number four is Azerbaijan, with production only slightly less than that of the United States, and expanding at the rate of some thousands of hectares per year. Number five is Georgia, also nearly equal to the United States; in fact, these positions can swap from year to year.

China is now number six and Iran number seven, not far behind Georgia. Both are increasing plantings.

Ten years ago, discussions of world hazel production put Spain at number four and France at number five. Now they are eight and nine, and it seems unlikely they will recover their ranks.

The real story extends far beyond the top 10, however. Hazels are now also commercially produced in Ukraine, Chile, Greece, and Australia, with very ambitious young plantings in South Africa, New Zealand, and more. The traditional industries in more than a dozen countries, including Portugal, Tunisia, Switzerland, Bosnia, and Russia, are now expanding. There's money to be made, and borders that were closed 20 years ago are now open. At the time of this writing, there is a great deal of risk capital available for large enterprises. Many of these new plantings have heavy investor support and are starting off with large "state-of-the-art" plantings and facilities—which means very low genetic diversity, relying only on "the best" hazel cultivars. Some are being attempted in the Southern Hemisphere. So far, these are all intending to simply exploit traditional established markets—more Nutella, more Godiva chocolate—and are investment- and profit-driven. One nontraditional but well-funded planting effort is going on in Bhutan. Their website says they've planted over two million trees—tissue culture clones. This project is not corporate-profit-driven (although there is a corporation), but is primarily aimed at social and environmental benefit.

I think the country to watch is China. The Chinese have been producing and breeding their own modern hazels for at least 25 years, that I know of, and traditional hazel culture is thousands of years old there. Demand for hazelnuts is up greatly with increased prosperity, and it is possible China could quickly become a major player in the international markets. In 1989, I took some of our Badgersett hazels to China (with full official quarantine) for their breeders to test and work with. China has abundant diversity of growing regions to allow them to use existing hazel types. Neohybrids, with much greater cold resistance and genetic diversity, could prove adaptable to Korea and Japan, both countries where growers are highly skilled with tree crops.

What the Future Holds

The markets for hazelnuts are stable or even growing somewhat, due to increased prosperity in China and some segments of many other populations. The increasing production from traditional sources is not likely to overrun the simultaneously increasing demand for existing uses in the near future. However, we believe there is considerable likelihood that climate change will make some traditional growing regions unworkable or unprofitable for the hazel clones currently in production. Neither shifting production to different regions nor finding new cultivars is an easy process. Disrupted production is quite possible, as is delayed establishment of new production. Prices for hazelnuts are likely to go up; supply is likely to go down. Political instability in producing regions also contributes to the trend.

Another very important upward force on hazel prices—indeed, on prices for all nut crops—is the publication of several studies from the medical and health community that all reached the same conclusions: Not only are nuts (with their high oil and caloric content) not bad for you, they are unquestionably very good for you, and may extend your life span.[3] If such claims come from anyone selling a product, you can be sure they're pure malarkey, but these studies were from prestigious scholars and well received by the world medical research community. In fact, the nearly unanimous verdict was: Yes, this research was very well done; yes, the studies were long enough, and broad enough, to be valid; and yes, the conclusions are warranted.

All this was picked up by the world popular news media and distributed around the planet at maximum decibels: "Nuts will make you healthy! Nuts will make you live longer!" Sales increases on nuts began immediately, and in 2013 world prices for nuts increased from 10 to 25 percent, depending on species.

World prices for hazelnuts are going up and are likely to keep going up for the near future. Current supply is vulnerable to climate change, and may go down. Political instability in producing regions[4] is a factor and could be a continuing contributor to downward pressure on supply and simultaneous upward pressure on price.

And on top of all this good news, as I tell my students, "Anything you can do with a soybean, we can do with a hazelnut. And then some!" Indeed, the potential commodity uses of neohybrids are exciting and diverse—a subject discussed in detail in chapter 11.

CHAPTER 4

Preparing to Plant

A neohybrid hazel planting is a perennial ecosystem, not an annual field crop, which means that planning your planting is considerably more involved. Before you start planning, though, it's important to spend some time reflecting on your goals, which will help you decide whether you should plant hazels and, if so, what scale of planting you want to embark on. Your goals will also affect many of the spacing, genetics, and field layout decisions you must address before you start planting your baby hazel bushes in the field.

Once you've got a plan to start working with, you'll move on to preparing the ground and your plant material prior to actual planting time.

Deciding to Plant

The planting decision is about figuring out where you fall on a continuum that ranges from planting no bushes at all to planting a maximum-density machine-pick-only planting. We don't recommend either of these extremes, but there are plenty of choices between the two. It pays to learn about the range of choices first so you can ensure that your decision works for you.

If the idea of planting hazels is brand new in your mind, reading this book is a great first step in deciding what scale of production might be right. This is a crop that you—and whoever eventually takes over your land—will be dealing with for decades, possibly centuries. Attend field days or short courses about hazel production, too, and spend some time with local growers if you can.

If you're already pretty serious about growing hazels, it can be a good idea to put in a test planting of 50 to 200 plants as soon as you can. This allows for hands-on learning, and for making mistakes on a small scale. You can apply what you learn to later plantings as well as to future decisions about whether to expand your hazel planting, and how. Fifty plants is enough that you'll encounter some of the same issues involved with planting acres at a time, and 200 is few enough that your loss won't be too extreme if all your trees fall victim to deer, drought, or unskilled help. (Please note: "Unskilled help" is *not* the same as "help that realizes itself to be unskilled.")

In order to establish a hazel planting, you need:

- A pioneering spirit.
- Somewhere to plant them.
- A way to prepare the land and plant it.
- The ability to care for and feed the hazels during their establishment years (until they're five to seven years old).
- The capacity for substantial hard work.
- Patience.

Is Woody Agriculture Right for Me?

This question is very important if you're thinking about growing hazels as a commercial or subsistence crop. If you're thinking about growing hazels for wildlife management or as a non-staple food, then it's less critical.

Hazel farming, and woody agriculture in general, involve paradigm shifts. They are fundamentally different from modern annual agriculture and require different behavior from the farmer. Woody perennial plants are not at all like annual plants. Even if you're an experienced farmer or gardener, if you treat hazels like a row crop, your planting will probably fail.

Two fundamental differences are the annual work schedule and the degree of urgency associated with what you need to do on any given day. Many tasks associated with caring for hazels need to be done within a certain time frame, but that time frame might be "next month" or it might be "sometime before May," not the "right now!" that is the usual situation for annual crops. A capacity for long-range planning and follow-up is important, because although the time frame may be long, you still need to accomplish tasks within it or you'll end up with bad results—results that cost you 5 to 10 times more work or significantly more dollars than if you'd done the job on time. This particularly applies to many management tasks, including fertility management, woody weed control, and coppice. There will be occasional emergencies—especially associated with harvest pest management, seedling establishment, and harvest—but far fewer of them per year than is the case with annual agriculture.

Another big difference in a woody agriculture system is pest management. We use and strongly recommend "ecosystem pest management," which requires a very different mind-set from the modern, supposedly surgical or carpet-bomb pest eradication. "What can I spray for this?" is a question you should remove from your mind. Instead, ask "How can I discourage this?" "How can I attract pest predators?" and "Does this ugly leaf actually matter?" If you are starting with a near-sterile agricultural landscape, establishing a hazel-based ecology takes time. In order to succeed, you need to be willing to grit your teeth and let new pests do some damage when they show up (see chapter 6). Can you trust

TABLE 4-1. Comparison of Mental Attitude Requirements in Different Types of Agricultural Systems

	Annual Agriculture	Woody Agriculture
Schedule outlook	Do it today!	Do it sometime this month.
Management style	I can deal with continual emergencies.	I can trust myself not to ignore the non-urgent stuff.
Planning perspective	I can imagine what the field will look like next week/month.	I can imagine what the field will look like in the next year/decade.

yourself to *never* go out and spray for weevils, eastern filbert blight, or tent caterpillars?

Do I Have Somewhere to Plant?

Finding a place to establish a long-lived perennial crop of hazel bushes is a more complex consideration than locating space for a garden or even a cornfield. If you want or need your planting to make economic sense, you need to be sure you'll reap the benefits of the investment you put in during the establishment years. In the United States, this usually means ownership of the land, but a long-term lease or other similarly crafted agreement could work. Land tenure should be carefully considered. Will you have access to the land for as long as you need? Do you have somebody in mind to take over when you are done with farming? If you plant there, will you be able to keep your crop protected from pests and have access to the land for harvest in August and September?

Can I Establish a Planting Successfully?

In their first year, newly planted hazel bushes can need care weekly—or more often than that. You, or a seriously dependable and attentive proxy, will need to provide this care. If you plant carefully and take some medium-range measures to address weed, water, and fertility concerns, you may be able to skip the weekly care routine without having them all die. However, mortality will be much higher and growth will be much slower than in a more intensively managed situation. A scenario that calls for spending one week planting hazels, then leaving the young plants to fend for themselves and returning to harvest a crop six years later, is *not* realistic.

Can I Do All That Work?

If you are thinking about growing hazels as a crop, you are thinking about *farming*, and there is no escaping the reality that farming is risky, weather-dependent, and work-intensive. Established hazels can survive almost any weather short of a direct hit by a tornado—but what do you do when it decides to rain for the entire month of September? (Hint: You get wet.) The establishment years of a hazel field can be a case of long-delayed, and even unreliable, gratification. You will put in a lot of work and have little to show the bank for it in the first half decade or more.

You will also need to be able to shift gears when production starts, to deal with more regular-farmer-like concerns related to harvesting and marketing the crop.

Do I Have Enough Patience?

Patience is listed as a requirement because *woody perennials* means slow, and *sustainable* also means slow. Plant trees for the future. Plant a tree or bush now, and you'll get fruit, or nuts, or shade, anywhere from 5 to 20 years from now, if not longer. Most folks, especially those with experience, and hopefully you as well, understand that no matter what you do there is going to be a waiting period between input and output for a tree or bush. You'll experience the joy of planting and helping things grow, but also have to cope with relatively low levels of monetary and caloric payback for many years following planting. Managing some intercrop in the meantime, between your young hazels, can certainly produce income, but will typically require a good deal of work and very different skills. We particularly like pastured poultry between hazels; but that is a full-time job if you want it to be profitable.

Slowness in the sense of sustainability is a different concept. There are some links between the two, and often people interested in sustainability will also be interested in slowness and its benefits. The Slow Food and Slow Money movements are examples. We do, however, need to provide some antidote to the ideas that often surround sustainable development. Governments and corporations often sell this term as "we can keep up this economic expansion in a way that will go on forever." But endless growth is never actually attainable. In order to be truly sustainable, the development has to truly be *growing*, rather than *mining*. This is a much more rigorous requirement. The lion's share of modern agriculture is literally mining the soil, not to mention the associated actual mining for fossil fuels and nutrients. It *is* possible to grow/gain wealth/improve by growing things *without* mining, but it is much slower than mining. It is, possibly, sustainable indefinitely.

If all this talk about slowness worries you, start your test planting as soon as you can. It might reduce the amount of patience you'll need to cultivate on the road to a successful long-term planting.

Defining Planting Goals

Defining your goals is key to many of the other decisions you'll make in planning a planting. Why are you considering planting hazels? Your goals may be simple or complex in many directions, as shown in figure 4-1.

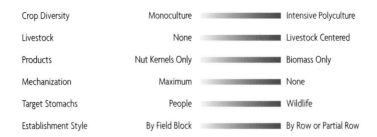

FIGURE 4-1. Several dimensions that can be used in defining planting goals. Where you aim on these spectra will affect planting and management decisions profoundly.

Crop Diversity

At one end of this spectrum is what we call the "coal-fired" orchard-style planting. Only hazels are allowed in the field, and everything else (at harvest, even grass) is kept out. It is very hard, even impossible, to do this sustainably. At the other end of the scale is a no-holds-barred perennial polyculture with multiple scales of production, with twenty to fifty different crop species intermingled, and perhaps many more "companion" plants. While this model has limitations for large-scale food production, it can be a very stable, productive, and fun way to produce food for yourself and those around you.

At this point, what we've found works well for us is blocks of hazels, perhaps 10 or 15 rows 200 to 300 feet long, with grass between hazel rows. The blocks are broken up by a tall tree row or rows between blocks. We plant to assure diversity in the fields, both species and physical, which enriches the environment by providing different microclimates for birds and insects. We have a row of pines between hazel blocks and are now planting single rows of hickory/pecan neohybrids the same way. We plant blocks of chestnuts among the hazels, but find the two do not get along in close proximity; the hickory/pecans like the hazels much better.

Livestock

Grass of one sort or another will be a part of any sustainable hazel planting. Mowing will control grass and weeds, but it takes time, money, and carbon fuel; it's better to do something more productive with your grass if you can. You can run poultry in most field plans with little or no modification of the plan—free range and poultry tractors don't require changes to row size. Chickens and guinea fowl will fill their crops with bugs and grass a couple of times over each day, reducing those pests while simultaneously harvesting their nutrients. Geese do, too, but they can also eat all the hazelnuts within reach; we've only tried that once. So far turkeys don't seem to eat the hazels off the bush much, but that might change with more experience; they're certainly capable of swallowing and digesting the nuts.

Larger grazers such as sheep require more advance planning. You'll need to figure out issues such as where to run the livestock fencing and how to prevent the animals from eating the bushes. Young plants may die or recent coppice may be damaged by being chewed on by small sheep like Babydolls, but mature bushes can withstand some browsing.

What sheep eat varies substantially by age and breed. Some growers have reported grazing feeder lambs among hazels without much damage to the bushes as long as they had enough grass. At Badgersett, we graze Icelandic sheep, but they have the voracious appetite of goats and will massacre hazels when the plants are leafed out. They don't seem to eat the bushes when they're leafless. The sheep even break out some deadwood, essentially doing some pruning for us, which allows us to wait longer between coppicings and makes the task itself easier.

Rotational grazing of cattle between rows of bushes might work as long as the cattle can't reach the bushes; this is something we have not yet tried. Horses work well in hazels, if well tended. When they have grass to eat, horses will not touch the hazels. But if left in a paddock until the good graze is gone, they will start eating hazel leaves. In winter, the horses won't eat dormant hazel twigs unless they're truly hungry.

Goats with hazel bushes is probably a ridiculous idea.

Target Products and Markets

What is it you are aiming to get out of this land and crop? Food? Money? Fame? Funny looks from your neighbors? More to the point, how do you want to market what you grow: at local food co-ops and other local markets, or do you want to make and sell value-added food products? Maybe you like the idea of on-farm sales, including pick-your-own? If you want to get money out of it, it's important to plant enough for a production level appropriate for your desired market.

Growers ask us about whether hazels can be grown organically. The answer is yes, and there are a few certified organic hazel growers in the United States, with more on the way, mostly in the tree-style Pacific Northwest culture. There are plenty of organic hazel growers in Turkey and some in Europe, but it's easier for them to achieve certification because they are very old hazel plantings in traditional culture, where sprays and synthetic fertilizers have not ever been used. Certified organic hazelnuts will bring you a higher price per pound for your product. Managing pests organically is certainly doable; our experience shows that spraying nothing is a perfectly good option, and it's what we do. One thing growers have found difficult is fertilizing the plants sufficiently using only organic fertilizer. Organic fertilizer is expensive, is usually low in nitrogen, and tends to be slow moving in the soil. The neohybrid hazels we now sell include high producers that do need fertilizer at a high rate in the first 10 to 15 years in order to reach full potential in that time—more than organic methods can easily provide. Some small producers and permaculturists are successful in avoiding all synthetic fertilizer. It does tend to mean longer times for the plants to come into production (two to three years longer). It is also more work, and yields are probably lower during the first 10 to 15 years. We currently believe the fertilizer requirements drop somewhat after age 15 to 20 if plantings receive good care; it is likely that attaining equivalent yields with organic fertilizers would eventually become less difficult.

One of the strongest arguments against using mineral fertilizers is that they can be very water soluble and mobile in the soil, and tend to easily wash away into rivers and aquifers. That is certainly true for row crops, but we've tested it multiple times in the hazels, and it just does not happen; the huge 365-days-a-year hazel root systems catch every molecule we apply.

Badgersett is not organic or certified, though our only non-organic practice is the application of mineral fertilizer, and we use no pesticides whatsoever. Because we and our practices are known to our chef and wholesale customers personally, they are quite happy both to allow us to call our produce "Beyond Organic!" and "natural." And pay us a premium price. Getting certified is itself expensive, and the record keeping required can be a pain. We're happy as we are.

Nuts aren't the only product you can market when you grow hazels. Perhaps your primary interest in hazels is to produce biomass, biochar, biodiesel, or pulp. It is possible to do a biomass-only planting, but you can't do a nuts-only planting. Wood and nutshell are inevitable co-products of hazelnut farming.

Mechanization

If you'll have a multiple-acre planting, consider what level of mechanization you may want or need for planting, coppicing, and harvest. Do you have a lot of family, migrant labor, or other people power available? Even if you do, keep in mind that planting, maintenance, and harvest of hazels are skilled jobs that require consistent quality control. It's more involved than most vegetable production because of the multiyear impacts of actions, and different from most orchard production in that neohybrid hazels are new, seedling plantings requiring different quality-control attention, and there are nearly no workers already skilled in these jobs. Do any of your neighbors grow hazels or plan to plant them? If so, perhaps you can (carefully) share equipment or the cost of custom harvesting.

Feeding Wildlife

Ecosystem pest management (see chapter 6) always involves wildlife of some sort. But field layout and other management practices can be very different depending on whether the goal is to produce nuts for people to eat or nuts for wildlife. This book is about growing hazels to feed people or livestock. If your wish is to create wildlife habitat—especially for squirrels and blue jays—read the field layout and animal pest management recommendations in this book, and then do the opposite.

 ## Site Requirements

Finding a site where hazels will grow isn't too hard. Getting them established is often more difficult, and getting nuts out of them requires attention to the limitations and advantages of your site. Here we discuss this in terms of landscape: the lay of the land and its surrounding habitats, climate, and soil conditions.

Landscape

In the North American wild, we've seen hazels growing in nearly any site and soil type imaginable: fields, forests, tops of mountains, hillsides, valley bottoms, stream-sides, fens, and at the edge of floating bogs. Here around Badgersett they usually establish most easily in shallow draws, where they have access to a little more fertility and water than the surrounding land. At the western edge of the native range, hazels tend to be riparian, with roots able to get to a nearby stream or pond. In the far south, they prefer mountaintops; see the notes on hazel distribution in chapter 1.

You'll need to be able to move through the field. Hazels will certainly grow on rough, steep land, but steeper slopes make care and harvest more difficult: The rougher and steeper it is, the harder harvest will be and the more you'll have to do by hand. You can find more variety with special-order machine harvesters, but most don't adjust well to a side hill of more than a 15 percent, or sometimes 20 percent, grade; and so far none are being built specifically for hazelnuts.

Your surrounding land can make a big difference in pest movement, bush growth, and human and machine access, thereby dictating which concerns will be most important in establishment and management. Nearby or adjacent woods can help speed up the development of a beneficial ecosystem, but they also provide habitat for squirrels, deer, blue jays, and other potential pests. You'll need to consider that when planning your planting layout (see *Field Layout*, on page 64). In most cases, for instance, planting in a small field surrounded by forest is not a good idea.

Your human neighbors are also an important factor. What are they planting and what sorts of pesticide drift can you expect? How much of the water in the water table the hazels' roots will be able to access comes from nearby agricultural land? There have been some growers' fields where we suspect very slow establishment is partly due to either herbicide drift or residual herbicide in the groundwater. Remember hazels' deep roots; what they find might not always be good!

Climate

The existing *Corylus avellana*–based hazel industry is centered in Mediterranean climates, but neohybrids are generally well adapted to a continental climate with greater hot and cold extremes. Neohybrid plants from a knowledgeable source with well-tested parent plants should be rock-solid hardy in USDA Zones 4 through 6. It's a reasonable risk to install large trial plantings in Zones 3 and 7. In Zones 2, 8, and 9, start with a small trial planting and see how the plants do, potentially expanding the planting after two to five years of bearing. Still, 100- and 1,000-year weather events might really hurt the bushes in these zones. There have been some reports of some neohybrids suffering frost damage after early bud break in maritime climates—on Washington Island in Lake Michigan and New Jersey, for example.

High winds can slow down establishment; if you've got sustained winds of 15 to 20 miles an hour for much of the growing season, wind protection may be necessary, particularly in drier near-prairie locations; you may also need a little more water, or more time for establishment. Wind protection in these cases can be anything from grass strips left unmowed to snow fences to individual windbreaker tents or tree tubes.

Hardiness Zones

The USDA hardiness zone map has changed; with the warmer climate, most places in the United States have changed half a zone or more in the past 20 years or so. At Badgersett we used to be at the cold end of Zone 4a, and some maps now put us near the warm edge of Zone 4b. On average we are certainly warmer, although we also just experienced the coldest winter in over 30 years, where we recorded 28°F below zero. The zone map and average yearly lowest temperature have changed, to be sure. The extreme excursions of temperature have stayed more or less the same, however, or even become more common. For this reason, our geographic recommendations for planting hazels haven't changed much, even though the zone map has. If you're using a recent hardiness zone map, it's a good idea to add a *b* to the zone recommendations we give.

Neohybrid bushes will thrive on 25 inches of rain per year or greater, access to a natural water table (such as along river bottoms), or irrigation. Rainfall at Badgersett has averaged near 28 inches per year recently. Soils with higher water-holding capacity will allow the hazels to do well with more erratic rainfall; soils that hold less water will need a more even distribution of precipitation (though not necessarily more), or access to the water table.

Soil

At Badgersett, we are fortunate to have generally fertile silt loams, with clay subsoil and limestone bedrock. But we have not yet found any physical soil type where hazels cannot succeed, from heavy clays to sands. As suggested above, the soil type does need to agree with the local water availability. Hybrid hazels can thrive with a soil pH between 6.0 and 7.5; they are usually seriously unhappy with pH under 5.5. Test plantings are a reasonable risk in soils with pH as high as 8.5.

If you want your hazels (or other trees and bushes, for that matter) to a produce a good crop, you will need *fertile* ground. It is, of course, possible to improve soil fertility over time, including after planting.

The soil's infiltration rate and water-holding capacity are certainly important; for clays in particular, these may take decades to fully develop in a woody ag field. Subsoiling during ground preparation to break up compaction and introduce deep air in the soil seems to always help root penetration and water security. We never plant without first subsoiling now.

For an initial evaluation of the suitability of your land to planting hazels, it can be useful to refer to the NRCS web soil survey (http://websoilsurvey.sc.egov.usda.gov/App/HomePage.htm). Take the time to learn the interface if you can; there is extensive information about soil and other site information there, including the National Commodity Crop Productivity Index. If you want a serious crop, an index of less than 0.5 to 0.6 (out of a maximum of 1.0) indicates you may have serious issues to contend with before you can hope for maximum production. In some cases, though, these problematic sites can be dealt with, or the particular problems don't matter as much for woody perennials as for annual crops. But you should be aware of what has brought the rating down, and have a plan for what you'll do to address it.

 ## What to Plant

Once you've decided *to* plant, and have some clear ideas about *why* you're putting in this planting, you'll need to figure out *what* to plant; in other words, where to buy planting stock, and what types of seedling genetics, and possibly clonal varieties, to choose. Keep in mind that you'll incur most of your establishment costs after the initial planting—adequate care for the first five to eight years is substantially more expensive than the plants themselves. In effect, this means that trying to economize by buying cheaper plants with less-good genetics will usually be quite a lot more expensive in the end. See the *Paying for It* section at the end of this chapter for details.

Genetics

Most fruit and nut crops are based on genetically identical plants: clones, of one kind or another. Cloning of trees is an ancient practice, normally accomplished by grafting or layering. All the apples you buy at the store are from grafted trees, for example. Commercial hazels in the Pacific Northwest have been primarily propagated by stooling or layering, though tissue-cultured hazels of this type are starting to become commercially available. Any clonal hazel you can buy today is extremely unlikely to survive outside the Pacific Northwest.

The great majority of neohybrid bush hazels planted to date are not clones, but grown from seed. This is partly due to the difficulty, expense, and time investment involved in cloning whole hazel plants. At this point, however, even if cloning became cheap, easy, and reliable we would still recommend primarily seedling plantings for the foreseeable future, because there are no thoroughly tested clones. Truly testing a new clone requires decades of observation; the very popular Honeycrisp apple, for example, was in testing for thirty years before the breeders released it to the public. Anything close to being a tested hazel clone is from a (much) earlier selection cycle compared with current seedlings available from Badgersett, and may be behind improvements in seedlings available from some other growers as well.

Recent advances in clonal production at Badgersett may make the sale of *tested* clonal material available within the next few years. Besides our own work on improving divisions, one of our partners is now working intensively on tissue culturing both Badgersett selections and his selections from his planting in New York.

And—never forget—the climate is changing. Even well-tested clones risk trouble in a changing climate; a seedling planting is very likely to have some plants that do well in the changed environment—this is how woody plants have always survived in the past.

The type of planting you intend will have an effect on the genetic mix you should pick. Essentially, *genetic mix* refers to the traits that have been selected for in the parents, and to what extent the progeny are likely to be variable. Choosing the genetic mix for your field can be a matter of considerable art.

TABLE 4-2. Respective Advantages of Hazel Clones versus Seedlings

Clones	Seedlings
Repeatable and predictable.	Only 60 percent will strongly resemble female parent; the rest will be worse or better.
A tested clone may have well-known responses to past weather and pest events.	Manage the planting to adapt to your location and climate changes.
Blocks of consistent plants for easier machining.	Manage your field for improvement to machinability over time.
Blocks of plants with the same weaknesses.	Diversity gives rise to resilience in extremes.

In Badgersett's current catalog, seedling options often have a kernel size class as well as some indication of whether they are long-tested (select) or recently developed (experimental) genetics. Most of the material Badgersett sells as "experimental" are good performers for which we don't yet have many years of data; many of these will become tomorrow's more advanced tested, and even select, material.

We are working on separating out traits related to machine picking and handpicking. What you select would depend on what type of operation you envision running. Here are some examples.

PICK-YOUR-OWN. One strategy would be to plant a mix that will give a range of nut sizes. A good mix might be 50 percent large, 30 to 40 percent medium, and 10 to 20 percent extra-large. This mix won't give you maximum yield per acre, but it will give you maximum customer satisfaction, because customers like finding and picking large hazelnuts. If you can plan for it, intersperse some of the trees with the genetics for extra-large nuts throughout the planting, so that your customers can find them here and there.

You'll probably want to choose handpicked parents if you can. For each size class, you might choose 50 percent elect/tested stock and 50 percent experimental stock. Or you could choose a higher percentage of experimental, then plant in at a higher density and thin the planting over time. Eventually, this should result in a higher-quality field.

WILDLIFE HABITAT. Choose a majority of stock classified as wildlife (high production with small nuts; some of these bushes hold the nuts on the bush for months), mixed with some standard-sized and some experimental large.

MACHINE-PICK PRODUCTION FIELD. Plant some combination of machine-picked early and machine-picked late, depending on what you're expecting your target markets to be and when you expect the machinery to be available. The neohybrids typically ripen from August 10 to September 20, and we now keep the early and late genetics separate. If you don't have specific needs in mind, we recommend half early and half late, planted with rows of the same type all together in the field in order to simplify and spread out the work of harvest. Also, as in most production plantings, plant 20 to 50 extra-large bushes, just so you can show people your big nuts. Seriously, there is something deeply, primally attractive about big tree nuts. A single huge nut gets a much bigger "Oooooh!" response than a whole handful of little ones, which may taste and crack much better. It complicates production, but size is great for marketing.

Whatever your mix of size classes is, we recommend that production-oriented plantings use a 70–30 to 50–50 mix of guaranteed–experimental germplasm when both are available.

Other suppliers don't offer the level of choices that Badgersett does. In most cases this means the planting stock they offer is a central subset of whatever selection cycle they obtained from Badgersett, with whatever overall selection (both intentional and inadvertent) they've been providing since then, often with local wild hazel pollen mixed in.

Hazel seed parents or clonal cultivars with strong local adaptations or test records for your location are unlikely to exist yet. Talk to suppliers before you buy from them. Ask questions about the seed parents such as: Has the bush been coppiced? Has it been through a test drought? Has it been thoroughly exposed to pests of interest? Does it have special pollination needs? Choose plants from somebody who can answer some of these questions; even their untested material is likely to be better than other growers'. If they cannot answer questions like that be very careful.

Seedling Type

Transplantation is an unnatural act. In the wild, woody plants are not in the habit of heaving up their roots and moving to a new site. And like most long-lived woody plants, hazels prefer to be in their final location as early in their life as possible. We discovered only about 8 years

"So You Live on a Nut Farm, Huh?"

As a grower of hazelnuts, you will need to contend with the deep truth that nuts are funny. *Nut* can refer to at least three things that definitely don't grow on trees, and the word for "nut" in other languages often has similar humorous uses. Your friends will poke fun at you about living on a nut farm, or being a nut nut. New acquaintances will occasionally make remarks about the size, placement, or number of your nuts. Nut crackers and nut husk removers will be discussed in a twisted light. And, so far, no newspaper editor has been able to resist putting a "nutty" pun into their headlines, regardless of how serious the writer was.

This is not for the faint of heart. You need to be able to take it! Be prepared. Our recommendation is to just embrace the nut humor, if you haven't already. Avoid telling folks you've heard that one a thousand times; they think they're being original. Pick out or develop your own favorite nut joke. Enjoy it! After all, nuts are funny.

ago that some neohybrid hazel lines just don't transplant. That's not an uncommon situation when dealing with wild genetics; we're working on getting those genetics entirely removed from our gene pool. Most of our seed now comes from plantings that were successfully transplanted, a great improvement.

The traditional method for transplanting woody plants and trees in quantity is to use dormant bare-root plants, but for many reasons we've developed a different system at Badgersett, using what we call "tubelings." Though it feels good to look at a large bush you've just planted, it's been proven many times that very young transplants usually take hold and grow faster than older stock, which may not tolerate the shock as well.

Tubelings

Tubelings are mini containerized seedlings approximately four months old when planted into their permanent location. Tubelings have multiple advantages over bare-root dormant plants. One, shared with many other containerized trees, is a much larger planting window. With dormant bare-root stock, the planting window is typically only two to three weeks long. If cold, rainy conditions keep you out of the field during that window, you've missed your chance to establish a new planting for a whole year. Tubelings can be planted from May 15 to July 30 in Zone 5 and colder, and until September 1 in Zone 6 and warmer. The fact that tubelings can be planted in May, June, July, and August is liberating. You can arrange to plant at a time when the ground is ready, the necessary labor is available, and other urgent needs are not conflicting. All this means that planting can be done at a lower cost and a greater probability of success.

Another difference is size. Dormant bare-root stock is many times larger than tubelings are, requiring more work and digging to transplant and—when large numbers are shipped—much bigger trucks to carry the same numbers. The bigger plants are also much more likely to experience transplant shock, and grow very slowly for up to three years.

Raising bare-root stock has other risks as well. Hazels typically leaf out with the very earliest trees. Very quickly in the spring, they are not dormant anymore and not suitable for treatment as bare-root dormant; experience has shown us that too often, hazels grown in the field will be leafing out before the ground is workable, which means that digging them for transplant in the spring is extremely risky. Fall digging requires refrigerated climate-controlled storage over the winter; our consistent experience with experimental trials is that hazelnuts do worse than most other plants in such storage.

Wild or unselected hazel seed is cheap; nurseries can afford to have animals eat most of it. Selected hybrid hazel seed from tested plants, identity preserved, is perforce expensive, and feeding 85 percent of it (a

FIGURE 4-2. Badgersett tubeling hazels ready to plant into field. Treat them like bare-root dormant stock and they will die. Treat them like tubelings and your success rate can be 80 to 90 percent.

real number from years of trying) to the mice, raccoons, and woodpeckers has not been economic.

The tubelings are therefore much cheaper to grow, protected by concrete and glass, and get into the ground than bare-root dormant plants can be for high-value rare nuts, which is what we have now. They are still not cheap, but we work on getting the price down every year.

Tubelings provide the most recent and advanced genetics available; this year we will put plants in the field from last year's harvest, instead of the one- or two-year delay (at least) involved in bare-root dormant plants. At this point in the development of neohybrid hazels, that is still a

FIGURE 4-3. These healthy young hazels, in their fourth leaf, were tubelings planted by machine—in heavy clay soil, in August, in a drought—and treated mostly according to the directions. More fertilizer would have been a good idea.

very substantial difference. It's like getting a current-model car versus one from the mid-1970s.

At Badgersett, we sometimes also produce "bare-root dormant tubelings." These are 8- to 10-month-old tubelings that have been over-wintered in the greenhouse and pulled while dormant. Although hazels tend to die if kept in dark refrigeration for several months, they not only tolerate being kept in a warm greenhouse over the winter, but actually thrive. This is not the usual expectation. Many plants kept in continuously warm circumstances will break dormancy and start growing—a bad thing if you want to transplant in early spring. The hazels, however, as part of their extreme cold hardiness, truly require a substantial number of cold hours before they will break dormancy. Kept warm, the tubelings do not meet their cold requirement; consequently they stay dormant, and happy, sitting in the sun in the greenhouse. In February we pull them out of their tubes and put them into cold storage for a month, which does not kill them, but does supply their cold requirement.

This kind of planting stock is available in quantity in some years, allowing plain old-fashioned early-spring plantings. They can be shipped, handled, and planted just like any bare-root trees.

Bare-Root Crowns

Larger or older transplants—for instance, four-year-old hazel plants—are not usually available at present. There is demand for them, particularly from homeowners interested in a few plants for their yard, but Badgersett has not had the resources to expand into this kind of nursery operation. This could be another niche opportunity for existing nurseries. Badgersett expects to continue to focus on producing mass numbers of plants, as inexpensively as they can be delivered to growers intending to get into serious nut production. Older transplants have uses there, also—for example, to fill in holes in an older planting—but the costs of such plants will always be very high in comparison with tubelings or tissue culture starts.

Hazel crowns are an alternative transplantation type that Badgersett has experimented with for several years. These are hazels that are three years old or more, but their handling is very different from standard transplant methods. Only the core of the root system is dug, anytime when the ground is not frozen; then their tops are removed *entirely*, 100 percent. This leaves just the root system. We have planted out these crowns with uniform success in mid-July, and since there are no leaves to support, the root system is not under drying stress even if it isn't irrigated. The plant will then send up shoots and put out leaves, but only as the root system is prepared to support them. This kind of regrowth from a total loss of the top is intrinsic in the genetics of the American native hazels, which are adapted to ecologies that include frequent fires. This system was investigated as a possible method

FIGURE 4-4. Bare-root crown survival in a Conservation Reserve Program (CRP) planting; the surrounding field was planted with tubelings at the same time. The field was allowed to be overrun by pocket gophers, and they wiped out the tubelings—but the crowns survived.

for establishing hazel plantings in very difficult circumstances, such as in dry sands, in ground subject to frequent flooding, or in situations with extreme animal pressure. The older root system contains very substantial resources and energy reserves to help the roots get reestablished, and the top, being entirely new, grows uniformly and without stress. Tests of the process were very successful; however, at present nursery stock of this type is not being produced, simply because Badgersett does not have the equipment to do it in an economic fashion. Digging three-year-old and older hazel root systems is a job that can require fairly powerful machinery—the roots are large and deep, and hand labor is not an option for a business unless you have a lot of really beefy employees.

Direct Seeding

We're often asked, "Why not just plant hazel seeds in the ground where you want them to grow?" We've tried this, in many different ways. The bigger the planting, the worse it works.

A fact of life hazel growers must contend with is that hazelnuts are *delicious.* They are not only the subject of cravings by humans, but also dramatically more attractive to wildlife, of all sorts, than beginning growers can believe—vastly more attractive than corn or wheat seed is. Mammals and birds will work 24 hours a day to get at a supply of hazelnuts, and they will succeed, despite all efforts to stop them. Usually they succeed in the first 24 hours even if the nuts are underground. If there are a lot of nuts, they will notify their friends and relatives to come feast, too.

What is a nut for? A nut's purpose is to be attractive to vertebrates, so that they will move the nut around and help with seed dispersal. A nut's primary purpose is generally speaking *not* to provide lots of food to the new seedling—plenty of plants grow like mad and produce huge trees from teensy tiny seeds.

Directly planting your hazel seeds out in the field is certainly an enticing idea. Planting seeds is a lot less labor than planting growing plants, and you avoid the whole set of problems a plant is exposed to at transplant. It would be nice if it worked, and we do still experiment with new methods, but so far direct seeding of neohybrid seed has *never* made anyone happy on either side of the transaction. Overall success is so abysmal that we still strongly recommend against it.

That being said, increased production of machine-harvested seed has changed things such that this might be worth a try; we are now selling seed as a matter of course, but mostly to folks with greenhouses. *Incredibly* intensive pest management during the first two to four months is likely to be necessary.

Hybrid nut seed is expensive. There is no resemblance to the cost of, say, wild black walnut seed. Until very recently when we began choosing some machine-picked seed to plant for enhancing machinability, all seed was hand-harvested, and identity of seed parentage was maintained through harvest, cleaning, storage, planting, growing, and shipping. Then add in 30 years of testing and the cost of controlling the pollination process. Random seed or seedlings, from random, untested hybrids, may look cheap, but when you get useless plants 10 years down the road, it turns out to be very expensive indeed.

Field Layout

Your hazels, once established, are likely to be there for a long time. Our best informed guess is that in the wild, once they've made it to an age of 10 they have a good chance of surviving from 500 to over 1,000 years. A bush planted in the wrong place could be hundreds of times more work over its lifetime than one planted in the right place. Taking a little extra time to figure it out isn't a bad idea.

For *any* hazel field layout, there will be several important considerations to take into account, and a few elements that are likely to be shared between plantings where nut harvest is a priority.

- A field boundary of some sort, whether imaginary or not.
- One or more zones of hazel bushes, each with a particular row spacing and orientation of rows in relation to the landscape and the sun.
- Barriers of various types to protect the hazel zone from nut thieving pests, drifting herbicides, and the like.
- Habitat islands for lending stability to ecosystem pest management.
- Raptor roosts.

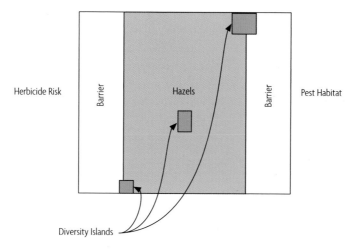

FIGURE 4-5. This abstract representation of a typical field layout shows landscape risks to production, barriers for those risks, and biodiversity islands for more effective ecosystem pest management.

Different risks require different barriers. Nut-thieving squirrels and blue jays will be slowed down if they must spend the extra effort to cross 100 feet of open ground safely, particularly if you install raptor roosts to increase the likelihood they'll be eaten on the way. Open ground doesn't provide a barrier to herbicide drift, but you can mitigate herbicide drift by installing a fencerow-style windbreak 15 or more feet tall. Keep in mind, though, that windbreaks can provide great cover for blue jays and red squirrels, becoming the opposite of a barrier for those risks. Human disruption should also be considered—a field full of nuts at harvest time is a nice place for the local kids to have a party; reducing road access and places to hide can help address that issue. (We've had it happen.)

Plant Spacing

What's the ideal spacing for rows of corn? How about distance between seeds within the row? Farmers, agricultural scientists, and agribusiness have been putting a lot of effort into maximizing production of maize for a long time, but no one has yet found the definitive answer. Certain limits and guidelines seem to apply under most conditions, but it depends.

The same can be said for hazels. Two parameters to consider are spacing at planting and final spacing. These are *not* necessarily the same thing, because you will lose bushes to predation or other causes, and you will probably want to remove bushes that aren't performing well. It may be tempting to put in your planting at the final spacing, then plan to replant as mortality occurs. There are pitfalls to this, however. (See below, and *Year 2* in chapter 5.)

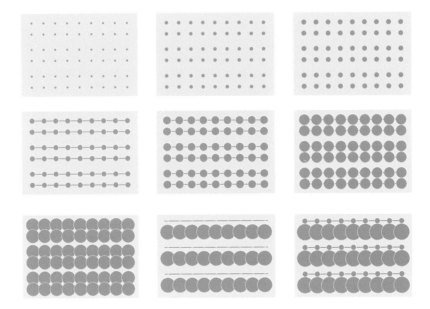

FIGURE 4-6. An idealized progression of growth for the first nine years of hazels spaced at 6 feet within rows, alternating 10 and 15 feet between rows.

In terms of what final spacing to aim for, 6 feet within rows appears to be a good final spacing, but 8 feet or more might be preferable for some types of handpicking. The highest-producing rows we've harvested so far are actually at a spacing of 3 feet or even 18 inches between plants.

Let's look at how the establishment years progress for two separate options: 6-foot initial spacing, and 3-foot initial spacing.

In the first year, we'll expect 80 percent survival. Further mortality will occur in the following years. Sometimes this will happen in blocks, due to concentrations of pests (particularly a gopher or locally marauding rabbit or deer). Figure 4-7 shows initial planting, mortality the second year, and survival expectations for year 4 in the two plantings. In the denser planting, this can be a good time to transplant from areas of denser survival to fill in holes where plants have died. Note that this figure shows expected survival, but keep in mind that some of these seedlings will not be great producers. About 40 percent you will probably want to cull. (It's often a good idea to coppice before culling, since some plants do kick into better production only after coppice. You could put plants you want to cull through coppice on a plant-by-plant basis in order to get them to either production or culling more quickly, rather than waiting to evaluate them at the first regularly scheduled coppice for the field.) Figure 4-8 shows the expected distribution of plants you'll actually want to keep for production. It should also be clear which of these options is more likely to actually lead to a harvest that could be profitable.

Distance between rows is another consideration. It's reasonable to expect neohybrid hazels to grow 8 to 12 feet in diameter, at least in the first few coppice cycles. Some plants will grow wider at the base, and post-coppice management will be necessary to prevent the bushes from growing too far out of the row. If you plan to use blueberry-picking machinery for harvest, a 10-foot on-center row spacing is minimum. For hand harvest a minimum of 12 feet on-center is recommended; it will feel too big for up to the first 10 years, but by year 20, a 10-foot aisle is uncomfortably tight for hand harvest. More space than this is needed for access for fertilizer application and harvest. In most cases we recommend alternating aisle width between 16 and 18 feet for this access, and 10 to 12 feet for minimum harvestable spacing. Spacing a field at all 10-foot rows will leave you with very little access to the field; it will be too tight for standard machinery (though highboy tractors could make it as long as you're not too overgrown), and the grown-up planting will essentially be too tight to walk through as well. It might be more manageable on a short (eight-year or less) coppice cycle, but to date no field planted at such tight spacing has reached maturity, so there's no track record to report.

And here's yet another consideration about row placement: Should they go north–south or east–west? We don't have definitive data on this, but some results from other bushes indicate higher production from rows going from north–south. Choose what fits your field best, including contours if the ground is steep enough. Under

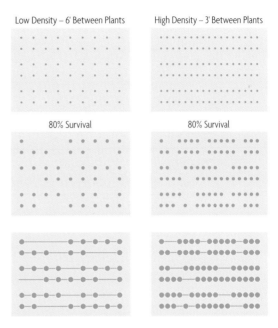

FIGURE 4-7. A more realistic progression of hazel fields planted at 6 feet between plants (left) and 3 feet (right). Eighty percent survival (acceptable) is shown at year 2, with further mortality by year 4, which is spotty due to concentration of pest damage in some areas.

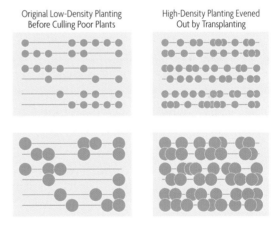

FIGURE 4-8. Continued progression from the previous figure at years 5 and 8. High-density planting at year 5 shows a more even distribution of plants due to transplantation from denser areas in the field; both plantings show a little additional mortality and some culling by year 8, representing the expectation that 60 percent of the seedlings will be acceptable for maximum production.

some circumstances you might even consider planting rows parallel to the direction of the prevailing wind when some windborne pest or disease is a local threat (see chapter 6 on, for instance, anthracnose).

 ## Experimentation

While you're planning field layout, it's a good time to think about the experiments you're likely to want to run in your field. Experimentation is critical because there is plenty more to be learned about growing hybrid hazels, and particularly about growing them profitably. Even if you don't want to experiment, it is extremely likely that at some point you'll come up with your own original concept for some aspect of managing your hazels, such as a better way to plant or fertilize or harvest. There's some chance that you'll be right, but the only way to know for sure is to set up an experiment to test your treatment.

"Doing an experiment" need not be difficult—it just requires some planning, attention, and observation. Here's an example of a very simple experiment: Fertilize half of your plants according to the instructions in chapter 5, and the other half according to some other scheme. Then observe what happens over the next few years. In this case, the half treated the standard way acts as a *control*. Comparison with the control is what allows you to determine whether your treatment produced results that were actually different (better or worse) than standard practice. Without a control for comparison, it is impossible to say with real confidence to what extent any improvement or degradation of the plants was due to the change in fertilizing practices.

Associated with this is the concept of confounding variables. In the field these often consist of differences in soil fertility, pest incidence, or weather pattern. In order to make your experiment most useful, you'll want to account for as many of these as you can—and experimental layout in the field can make a big difference. Let's consider the fertilizer experiment. Applying one type to half of your plants and the other to the other half can make sense, but consider which plants will receive which treatment. For example, does part of your field have better water-holding capacity than the other? If so, make sure that the fertilizer treatments are split as equally as possible over the two different areas. You don't want to apply the experimental fertilizer treatment only in the area that has the superior water-holding capacity, for example. Past treatment of the field is important to consider, too. Perhaps half the planting was a cornfield, but the other half is land that was in a conservation program.

Rough experimental setup can be fine—and is probably the best plan for many experimental desires. In most cases, you're looking for big, easily noticeable effects rather than marginal gains. This means just one control group and one experimental treatment can be fine, and measurements and observations can be coarse. Noting "about twice as much fertilizer" or "a shovelful" or "doesn't look like it made much difference over the first three years" can be more helpful than spending the time to measure values precisely or to lay out meticulously randomized test plots.

A little experimentation can be particularly useful in places that differ substantially from Badgersett Farm (as discussed in the *Site Requirements* section earlier in the chapter).

Ground Preparation

Proper ground preparation can make a big difference in how smoothly planting day, and establishment generally, goes. At the very least, you need to be able to get your plants into the ground without damaging the roots. Killing the sod is a really good idea, too. We do this by tilling. It's less expensive and more effective than using herbicides (we don't use herbicide at all in our hazel plantings anymore).

Subsoiling or "ripping"—using a single tine to break up compacted soils deeper than normal plowing would—is an excellent idea, especially in clays and old pasture. If possible, make three passes with a single-tooth subsoiler along the path of the row where you'll be planting the hazels, making one of these passes by driving in the opposite direction of the other two. If you are not tilling but are in conditions where subsoiling is important, use a pickax to break it up if possible, or use a deep (3- to 4-foot) auger to make the planting hole. If using an auger for planting, be sure to scarify the sides of the hole, because the auger action often glazes the sides of the hole, essentially creating a hard-shelled "pot" around the roots that will take years for them to penetrate.

Should you till in strips or the whole field? Strip tillage will slow erosion on places where you're putting contour or contour-tangent rows. It does mean you need to lay out and mark the rows on the field one more time, however, since you'll need to do it both before and after tillage. Also be aware that working in strips can sometimes result in shallow terraces, and these can make mowing or driving machines between the rows tricky or annoying.

For the silt loams at Badgersett Farm, the way we prepare the ground for planting depends on the size of the planting and the previous use of the land.

RECENT ROW CROP LAND. Two weeks before planting, we subsoil down the middle of the row, twice, in opposite directions, if the compaction seems serious. We then disk twice to break up clods and firm. Immediately before planting, we simultaneously blade and roll the row. The blade is set to level the soil with the adjacent untilled strips, and the

FIGURE 4-9. The beautiful strip tilling and cloudy weather provide perfect conditions for planting tubelings in this field at Badgersett Farm.

roller is pulled behind the blade, providing a firm, smooth track for the transplanting machine.

OLD HAY GROUND OR HEAVY SOD. Six months before planting, we plow either strips or the whole field, depending on needs. Land that has been out of production for some years can be very rough, and in need of leveling, often only possible via whole-field tillage. Preferably, the sods should have several months to break down before planting proceeds; the transplanting machines work badly and/or slowly in tough, chunky sod.

SMALL HAND PLANTINGS. We strip-cultivate or spot-cultivate so that an area about 3 feet in diameter is cleared of weeds. If planting in sod, sometimes the less disturbance the better; tillage brings more weed seeds to the surface to germinate. Undisturbed sod killed with herbicide or by covering with plastic/board/cardboard providing complete shade can act as a good mulch for as much as a year.

 Preparation of Tubelings

This section covers the treatment that tubelings need in order to toughen them up for the field, and how they need to be treated between the greenhouse and planting in the field. If you're planting bare-root dormant, hazels should be treated like any other bare-root dormant plant, but with the understanding that they'll probably leaf out earlier than most other bushes and trees.

Tubelings, like any greenhouse-grown plant, need to be acclimated to field conditions before outplant, or they will sunburn and then desiccate, resulting in damage or death of the plant. Usually this has been done in the greenhouse industry by taking plants through stages of decreasing shade upon removal from the greenhouse, but it turns out that decapitation works better for hazel tubelings.

Decapitation means we cut off the tops of plants before shipping, leaving only 6 to 8 inches or so of top left. It both toughens the plants and allows for fast field acclimation. Once the tubelings reach an appropriate height, we decapitate by the 98-plant frame, and once the secondary buds have broken there is a one- to two-week window for planting. Buds in the proper state will continue to grow through transplant, and will adapt immediately to the field conditions; plant before the buds have grown enough and the plant may just stop and sit there following transplant, which often leads to insufficient root growth and death. Additionally, decapitation toughens the stems and reduces the chance of wind damage. It also appears to reduce browsing on the newly planted tubelings, most likely due to a rather intense interplant signal to generate anti-browsing chemicals following the simultaneous decapitation of hundreds or thousands of plants in the greenhouse.

Plant as soon as you can once the secondary buds have grown ⅛ to ¼ inch. This will give them their best chance. Don't worry if you need to wait a few days before planting, but be aware that the longer you must wait, the more things can go wrong. Plants held a month or more past shipping may get potbound and may be slow to start growing again once planted; some have remained stuck for a year.

To hold hazel tubelings a day or so, be sure the roots in the tubes stay moist. If you must keep the plants more than a few days before planting, it won't hurt them, but they need attention. They will need to be watered moderately. Don't soak them; they don't need it, and it can make planting more difficult. Don't hold them in shade or refrigeration; sun is better for them. If you keep them in heavy shade, they will lose their adaptation to full sun. A refrigerator does this and also kills by dehydrating the plants.

Water the plants from the bottom up: Stand the tubes in a tub or other container with about 4 inches of water in the bottom. This is more certain than sprinkling water on them—spray from a hose always misses a tube or two, which will mean a dead plant. Dipping them for about one minute allows the plants to take up enough water to support them for a couple of days in most conditions. In hot windy weather, in the sun, they'll need water more often: Keep an eye on them! (See *Water* in chapter 5.)

Plants being held must be protected from animals before planting, since mice, squirrels, raccoons, and more can damage closely packed seedlings trying to get at the nuts. In some cases you can get away with leaving the plants unprotected in the sun during the day, and kept in the house away from pests at night.

6-Acre Example Planting Plan

To exemplify the recommendations in this chapter, here's a description of the field layout details for a 6-acre field (400 feet × 660 feet), with hazelnut production as the primary goal. This is a hypothetical planting in northeastern Iowa, with a fairly even southeast-facing slope of 15 percent.

Since the slope isn't too severe, east–west or north–south rows will work well—there's no need to follow the contour. East–west rows will reduce machine turnaround time, so that's what we'll choose. The slope is steep enough that careful attention will need to be paid to avoiding gully formation in the establishment year as described in chapter 5, but after that erosion should be a non-issue as long as we don't till it or keep it bare earth. We'll establish the first row of hazels on the north edge of the field 100 feet in from the woods. The 100 feet of open space will serve as a rodent and bird barrier. This space (totaling about an acre and a half) can be used for hay or grazing as long as it is well mowed at the start of the hazel harvest season. Similarly, a 30-foot border on two other sides of the field allows for both machine turnaround and pest management. We'll plant a row of pesticide-break trees along the road, including trees that will grow up and provide a windbreak quickly. These will also provide some habitat for ecosystem pest management support. If heavy and regular pesticide drift is expected, choose species that can withstand the

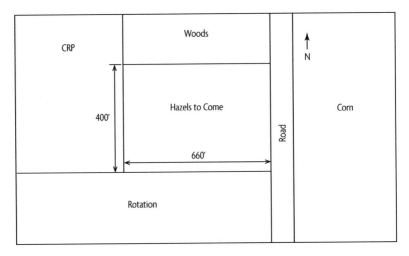

FIGURE 4-10. Our field has a small (2-acre) woods on the north border, a road to the east with neighbor's corn land across the road, Conservation Reserve Program (CRP) grassland to the west, and corn/beans/hay rotation land to the south (we own this land, but rent it out).

expected pesticide types; ask the local soil and water conservation district (SWCD) or extension forester for recommendations. In our case we don't expect heavy drift from the field of our conscientious neighbor, so we'll put in a mix: cottonwood to give us some quick height, black cherry for quick growth and bird-feeding fruit, white pine for a little conifer diversity, and an oak or two for good measure. Be careful, though—the cottonwood and black cherry will produce seed that will sprout as woody weeds among the hazels. Note that this break starts 150 feet from the woods; we'd really like this separation to reduce pest flow. We'll plant these trees at 10 feet from the east fence, and have the hazel rows start at 75 feet from that fence, giving an on-center spacing between the hazels and this crow habitat of 65 feet.

We're planning to machine-pick primarily, with an extended coppice cycle because of the nut production focus, so we'll space the rows at 10 and 18 feet alternately. Spacing within rows will be 3 feet. This gives us actual hazel growing space of (400 − 100 − 30 = 270 feet) × (660 − 75 − 30 = 555 feet), or about 3.4 acres. How many rows will fit? Every two rows of bushes (each double row) requires 10 + 18 = 28 feet, including the wide aisle for access. Divide 270 by 28 and the answer is 9 and change. The "change" gives us enough room for access on the southern edge, since we had extra space on the southern border to begin with. Ten double rows 10 feet wide and nine 18-foot-wide aisles between the double rows translates as (10 × 10 feet) + (9 × 18 feet) = 262 feet for the width of the planting overall, which fits nicely into the 270-foot width of the total growing space available. So this gives us 20 rows (10 double rows) 555 feet long; spacing at 3 feet between plants gives us an initial planting of 185 tubelings per row, expected to be thinned to about 92 (for a final in-row spacing of 6 feet).

We will install raptor roosts. (See figure 4-11 on page 74.) In order of most to least valuable, roosts should be put at spots 1 and 2 (this would be the minimum requirement), then 4 and 5, then 3, then 6 (in the middle somewhere) and 7, then 8. More than these would be fine, but minimize spots in the middle of the field where harvester/equipment negotiation will be difficult.

We won't put in a windbreak on the border with the CRP unless this is a high-wind area (one in which annual average wind speed at 30 meters is greater than 6 meters per second; this may not be necessary until the wind is greater than 6.5 m/s). The 30-meter wind resource map from the US Department of Energy (http://www.windpoweringamerica .gov/pdfs/wind_maps/us_windmap_30meters.pdf) lets us know that northeast Iowa is not generally what hazels consider a high-wind area. We're not concerned about storm wind damage so much as consistent, growth-slowing and moisture-sucking winds during the growing season.

Raptor Roosts

An effective raptor roost in a hazel field should reach at least 25 feet above the ground and have multiple roosting options, preferably from 10 or 15 feet all the way up to the top. They can be constructed from utility poles or, as we do, entire small trees. They need lightning protection (we just use 6-gauge copper grounding wire) and should be buried in a posthole to a depth of 6 feet. Tamp the soil very well (with a sledgehammer) when backfilling the hole, and stabilize it further with wooden fence posts driven in right against the roost on two or more sides.

Getting a 30-foot pole into a 6-foot hole in the ground is no joke. We recommend renting or hiring a utility truck or cherry-picker/crane to do this job. It can be done with a tractor, chain, rope, and a crew of five exceptionally strong and attentive people—but it's dangerous. If you want to try working it out that way, make sure you've got somebody experienced with logging or sailing in charge.

If this field were located in north*western* Iowa, then we'd most likely want a windbreak on the west field border, and maybe even a wind fence or two in the middle of the field during establishment.

With this field, setting up a mulch experiment would be useful, because the topsoil is good quality but not very deep. So you want to

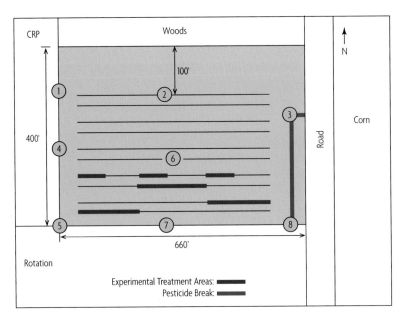

FIGURE 4-11. The plan for our 6-acre field, with measurements for row placement, includes numbered raptor roosts as the brown circles, and a layout for the experimental treatment areas that should avoid some confounding variables.

mulch, and let's say that wood chip mulch is available from a neighbor. Try mulching alternate half rows. You might want to try applying mulch at two different depths (see the figure 5-7 on page 98). Be sure you leave plenty of hazels unmulched as a control and always keep mulch 6 to 10 inches away from the hazel stem.

With your field layout decided, we recommend a gradual approach to your actual planting, unless you've already had experience managing hazels at a nearby site. In the first year, plant a single row of bushes, probably the southernmost; harvest the rest of the field for hay, or graze it if possible. Next year plant two or three more rows, and replant the first row if mortality is over 20 percent. In this field, you might only plant every other row to begin with, in order to leave more room for using hay or grazing as a purse nurse. The third year, plant up to 5 more rows—10 if you're really serious and have emergency help available for watering. Whether to install drip irrigation depends on how many plants you're planting in one year. Each year of planting, it can make sense to also replant the preceding year's trees if mortality is over 20 percent *and* the replanting won't screw up an experiment by confusing mortality or other data, or disturbing experimental treatments. This means you need to keep track of where replanting happens, and make sure that you maintain the different treatments between your mulch groups.

 Paying for It

Remember that bit about *sustainable* meaning "slow"? Well, another way to put that is "long-term," as in "You'll probably get your money back in the long term." Even when you're working with people who are very supportive of the ideas of ecological reconciliation and agricultural stability, this sentence can give you trouble with the bank, the spouse, and your own quality of life.

So let's look at some numbers and discuss strategies for not going broke during the establishment phase of a hazel planting. There are a lot of caveats here, because real data on people making money with neohybrid hazels are sparse to nonexistent yet. We'll try to make the assumptions for these numbers clear. Unless stated otherwise, we'll be using "good" or "excellent" care for the examples, because otherwise *long-term* can easily become "after you're dead, or when somebody starts taking care of the bushes."

One basic question is: How big does a field need to be to produce a satisfactory financial return? Essentially the more diverse your operation is, the smaller you are likely to be able to get away with, presuming you are actually selling or using all of your outputs. Also, the minimum field size for breaking even financially is likely to be smaller if you are marketing high-end value-added products. Our current estimate is that we could definitely make money on 4 acres of hazels grown for high-end local markets, but it'd be easier with 6. For commodity-type markets it might be closer to 15 or 20—a risky proposition at this point. Of course, *make money* means "make more than we spent," not "make enough to support a family of four."

A few aspiring hazel farmers have written formal business plans, and University of Wisconsin–Extension agriculture agent Jason Fischbach has put together an enterprise budget tool (http://midwesthazelnuts .org/assets/files/hazelnut%20enterprise%20budget_beta.xlsm), which is a good start. There are plenty of other tools and analyses for establishment of woody plants in a field; those closest to being useful models for hazel plantings are likely those for organic blueberries. There are some important things about hybrid hazel establishment generally missing from these tools that might not be obvious, however.

- Crop estimates are sometimes calculated as a direct function of plant density. However, with hazels, above a certain density (probably around 2 feet within rows and 10 feet between rows) maximum crops will not increase.
- Assumptions are for clonal field management style (or are based on that), rather than seedling field management style.
- Gradual onset of crop and switch-over from by-hand to mechanized planting and harvest are not accounted for.
- As in most cases for such tools, effects of care on how quickly production starts are not explicitly taken into account. So, for

Purse-Nurse Crops

In forestry and agronomy, a nurse crop is one planted to provide protection and perhaps nutrients to the final or primary crop, in order to achieve faster or more reliable establishment. In the establishment of a woody ag field, we sometimes use the term *purse-nurse crop* to refer to an intermediate crop or practice intended to provide cash flow from the field during establishment, thereby increasing the chances that you and your pocketbook will survive the establishment years. If you have enough resources to wait out the establishment years for your returns, that's great! Even if you do, though, it might be useful to try this kind of practice.

We currently put in most multiple-acre neo-hybrid hazel plantings as contiguous blocks of hazel rows, each block consisting of 10 to 20 rows. An alternative is a more gradual plan, where you plant only every other row to begin with, and continue harvesting the land in between rows either for current crops or other intercrops to financially ease the transition. Other types of purse-nurse "crops" can be pastured poultry (or other livestock with a bit more fencing), vegetable crops in the aisles, or even agritourism.

Remember: The most important element of a farm (without which there would be no farm) is *the farmer*. The farmer must be protected if the farmland is to be protected.

TABLE 4-3. Harvest Derating: Effects of Care and Coppice on Expected Hazelnut Harvest over Time

	Year 1	2	3	4	5	6	7	8	9	10	11	12	Full Potential per Acre
Good care	0	0	0	0.25	0.40	0.50	0.70	0.85	0.95	1	1	1	1,000
Excellent care	0	0	0	0.25	0.45	0.60	0.80	0.95	1	1	1	0.95	1,500
Marginal care	0	0	0	0	0.15	0.25	0.40	0.55	0.65	0.70	0.80	0.80	600
No care	0	0	0	0	0.10	0.15	0.20	0.25	0.30	0.30	0.35	0.30	200
Post-coppice good care	0	0	0.30	0.80	0.95	1	1	1	0.90	0.85	0.85	0.80	—

example, if you plan on not fertilizing or mowing, you would have to change the assumptions regarding length of time before a particular amount of harvest can be expected. (Some tools assume a specific time to first crop, and therefore don't allow you to make this very important adjustment.)

- Management costs are for standard orchard-style management, and ecosystem pest and disease management costs and benefits don't fit well into this model.

As a modeler of neuro-mechanical systems using robots, Brandon has extensive professional experience modeling complicated systems with lots of uncertainty in them. Unfortunately, the "simple" interaction diagram he came up with for the dynamics of the hazel field has 7 inputs, 11 variables defining the state of the field, and 5 outputs. This looks okay, until the 40 most important influences are drawn in. It's helpful if you're programming a model, but not particularly useful if you're looking at it for less than 20 minutes.

So, rather than turn this into a monograph on financial modeling of hybrid hazel growing, we'll go through a couple of budget examples based on the 6-acre plan in the preceding section. First let's address the important question of when to expect a crop, as seen in table 4-3. The table shows that harvest doesn't begin until year 4 or 5, and then the quantity of production and the speed of increase in production vary according to the care of the plants. Let's define what the care terms mean:

EXCELLENT CARE. Fertilized and watered to keep the plants continually green and growing, weed control three or more times yearly, browse and gopher control allowing less than 5 percent of plants to be damaged.

GOOD CARE. Fertilized nearly always on time, weed control twice yearly or more, wilting not allowed, browse and gopher control allowing less than 10 percent of plants to be damaged.

MARGINAL CARE. Timely fertilizer about half the time, weed control maybe only once a year, browse and gopher control only when it starts to look pretty bad.

NO CARE. Water and maybe a little fertilizer the first year; maybe one pass of weed control the first year, nothing after that.

The table shows the expected harvest as a percentage of the full-potential yearly harvest. For "good care," full potential is 1,000 pounds of dry whole nuts per acre (which is a conservative estimate). In year 4, the expected harvest is 25 percent of full potential, then 40 percent in year 5, increasing to full potential by year 10. Note that this table assumes a planting started from tubelings; bare-root dormant material will usually follow a similar or delayed-yield pattern even though the plants are larger to start with, because of the delaying effect of increased transplant shock. The bottom row of the table shows the rebound of harvest after coppice, which happens after the establishment period but is an important factor to consider in figuring costs and returns (thus we include it here). There will be two years with no production, then a year with 30 percent, followed by a rebound to near-full production.

The data on when to expect how much of a crop are necessary for any realistic projections of harvest, and therefore income; you can plug these numbers into table 4-5 in order to compare the expected results of various care regimes.

Because we'll be doing a gradual planting, and because it takes time for nut production to kick in, we need to calculate acreage and rows used in each activity for each year. Table 4-4 shows these numbers, described below.

ACRES TO PRE-PLANT. The acreage that will need the ground prepared for planting the following year.

ACRES TO PLANT/ROWS TO PLANT. Acreage/rows to be planted this year.

ACRES IN PLACE. Acreage of hazels that have been planted, following this year's planting.

ACRES TO HARVEST. Equivalent acres of hazels producing a crop, calculated here using the "good care" row in the harvest derating table, with the acreages planted in the preceding years. Note that this is a bit of a kludge as far as calculating harvest costs is concerned, since some harvest costs depend on physical area or row length harvested, rather than on nut crop density.

In table 4-5 we have a summary of establishment and management costs, and income, during the first 11 years of the 6-acre field described in

TABLE 4-4. Acreage and Rows for Cost and Harvest Calculations

Costs per Acre	Pre-Planting	Year 1	2	3	4	5	6	7	8	9	10	11
Acres to pre-plant	0.15	0.30	1	2	0							
Acres to plant	0	0.15	0.30	1	2							
Rows to plant	0	1	2	5	12							
Acres in place	0	0.15	0.45	1.45	3.45	3.45	3.45	3.45	3.45	3.45	3.45	3.45
Acres to harvest (equivalent)					0.04	0.14	0.45	1.16	1.64	2.10	2.69	3.10

the *6-Acre Example Planting Plan* above. Year 1 is the first year hazels are planted; note that there is some preparation work the year before. Here are definitions of the cost categories in the table.

GROUND PREP. Three passes each for acreage to be pre-planted and planted this year. Heavy sod may take a bit more, recent row-crop land a bit less.

PLANTING. Uses costs from table 4-6. No replanting is included, since with good care we expect less than 20 percent initial mortality, which should be fine with the 3-foot in-row spacing.

WATERING. Presumes three watering passes the first year, and one the second. This could vary widely and be 10 times as expensive or more in severe drought.

MOWING. Three passes with a tractor-driven flail/chopper each year.

NON-MOWING WEED CONTROL. Either three passes of establishment cultivation (usually with a tractor) or one pass of woody weed control (usually with a saw and shovel), depending on the age of the row. This one can become substantially more costly if you wait too long.

BROWSE CONTROL. Egg spray on plants one to three years old, three passes per year.

FERTILIZATION. Bulb planter application once shortly after planting and once at year 3; broadcast starting year 4.

PREDATOR SUPPORT. Labor for installing hawk roosts. Later on, management of pest movement corridors and barriers to allow greater predator access, including work with both a mower and a chain saw.

GOPHER/PEST CONTROL. Labor to scan, trap, and reduce pocket gophers; possibly also rabbits, ground squirrels, and other pests. If you start out with or develop a big infestation, this number can increase substantially (or you'll lose more plants).

HARVEST COST. Fairly good efficiency/management is presumed for production-oriented rather than data-oriented harvest. Hand-harvest years 4 through 6; machine-harvest thereafter.

TABLE 4-5. Income–Cost and Harvest Summary

	Pre-Planting	Year 1	2	3	4
Ground prep	$6	$17	$50	$116	$78
Planting		$730	$1,189	$3,964	$7,928
Watering		$9	$22	$68	$144
Mowing		$18	$53	$170	$404
Non-mowing weed control		$10	$21	$69	$138
Browse control		$16	$47	$152	$362
Fertilization		$50	$100	$300	$699
Predator support				$650	$650
Gopher/pest control				$270	$270
Harvest cost					$39
Miscellaneous labor/travel/expenses	$1,000	$750	$500	$500	$500
Total yearly expenses	$1,006	$1,600	$1,982	$6,259	$11,211
Cumulative expenses	$1,006	$2,606	$4,587	$10,846	$22,057
Nut harvest (pounds)					38
Purse-nurse income			$750	$750	$500
Hazel income					
Gross income	$0	$0	$750	$750	$500
Cumulative gross income	$0	$0	$750	$1,500	$2,000
Cumulative net income	-$1,006	-$2,606	-$3,837	-$9,346	-$20,057

MISCELLANEOUS LABOR/TRAVEL/EXPENSES. Additional field time looking at the plants, finding and debugging problems. Some learning curve costs are included here in the first few years and with establishment; some travel and education costs are also included.

TOTAL YEARLY EXPENSES. All the above expenses added up. This doesn't include landownership or rental cost, however.

CUMULATIVE EXPENSES. Sum of all the yearly expenses incurred on the planting up through the given year.

NUT HARVEST. Pounds of dry, in-shell nuts harvested; using acres to harvest (from table 4-4) and the full potential per acre number for good care.

PURSE-NURSE INCOME. Non-hazel net income off the whole 6 acres. This starts out as hay (the price of which fluctuates substantially) and transitions to pastured grazers. There are a multitude of possibilities for co-cropping and grazing, and the grazers should reduce mowing and fertilizer costs, which is not included in this analysis.

	5	6	7	8	9	10	11
	$0	$0	$0	$0	$0	$0	$0
	$41	$0	$0	$0	$0	$0	$0
	$404	$404	$404	$404	$404	$404	$404
	$12	$36	$116	$276	$276	$276	$276
	$347	$315	$210	$0	$0	$0	$0
	$388	$1,045	$1,059	$1,059	$1,059	$1,059	$1,059
		$500	$500	$500		$500	
	$270	$270	$270	$270	$270	$270	$270
	$140	$461	$526	$745	$954	$1,222	$1,411
	$250	$250	$250	$250	$250	$250	$250
	$1,851	$3,280	$3,334	$3,504	$3,213	$3,980	$3,669
	$23,908	$27,187	$30,522	$34,026	$37,239	$41,219	$44,889
	135	445	1,155	1,638	2,098	2,685	3,100
	$1,000	$1,500	$1,000	$1,000	$1,000	$200	$200
		$668	$1,733	$2,456	$3,146	$4,028	$4,650
	$1,000	$2,168	$2,733	$3,456	$4,146	$4,228	$4,850
	$3,000	$5,168	$7,900	$11,356	$15,503	$19,730	$24,580
	-$20,908	-$22,020	-$22,622	-$22,669	-$21,736	-$21,489	-$20,309

HAZEL INCOME. Calculated as $1.50 per pound for dry in-shell sold in bulk to a processor. If you husk and dry the nuts on the farm (which costs money), you'll be able to sell them to a distributor for a higher per-pound price. Processing costs are highly variable; this price incorporates a reasonable estimate of $1.50 per pound (see table 4-6).

GROSS INCOME. Sum of income streams for this year.

CUMULATIVE GROSS INCOME. Sum of all the gross income from this planting through the current year.

CUMULATIVE NET INCOME. Cumulative gross income minus cumulative expenses. This is how much money you've made on the planting so far.

Table 4-6 presents further detail of the cost calculations associated with the activities in table 4-5.

Rows:

TRACTOR. Cost of operating a utility tractor or harvester, minus fuel.

TABLE 4-6. Activity Costs during Establishment for 6-Acre Planting

	Ground Prep	Planting Year 1	Planting Year 2	Planting Year 3	Watering	Mowing	Establishment Weeding
Tractor	$5.00	$30.00	$30.00	$30.00		$15.00	$10.00
Fuel	$5.00	$10.00	$10.00	$10.00	$1.50	$10.00	$5.00
Consumables							
Equipment	$0.25	$4.00	$4.00	$4.00	$3.00	$4.00	$3.00
Labor	$2.67	$120.00	$120.00	$120.00	$16.00	$10.00	$5.00
Plants		$4,700.00	$3,800.00	$3,800.00			
Total	$12.92	$4,864.00	$3,964.00	$3,964.00	$20.50	$39.00	$23.00

All costs are expressed as dollars per acre except for Fertilizer, Establishment (dollars per plant) and Husking/Cleaning Cost (dollars per pound dry equivalent)

FUEL. Fuel for operating the tractor, truck, and/or harvester.

CONSUMABLES. Non-fuel supplies used in the activity; described in activity description.

EQUIPMENT. Cost to operate additional equipment for this task.

LABOR. Mostly calculated at $8 per hour; some higher-skilled jobs at up to $15. This includes your own labor, working on the assumption that your time is worth something elsewhere, and should be accounted for. This is a good idea for evaluating feasibility of a plan. A more fine-grained accounting for labor reducing the self-cost of the farm family labor could give rise to more accurate project costs, depending on how profits or wages are divided in your operation.

PLANTS. Cost of seedlings purchased; the price is higher the first year due to not reaching volume price breaks, which activate for the higher volumes in later years.

Columns:

GROUND PREP. One pass on 1 acre with a 90-horsepower tractor and 8-foot implement width; 20 minutes.

PLANTING YEARS 1–3. 1 acre; two hours using the same tractor, with transplanter.

WATERING. One pass on 1 acre with a tank on the back of a pickup truck; one person. This should take about two hours (longer in very dry conditions, though), not including time to fill the tank. Cost of the water itself is considered negligible.

MOWING. One pass on 1 acre, using a 6-foot flail/chopper mower and the tractor; one hour.

ESTABLISHMENT WEEDING. One pass on 1 acre, cultivation or mow-over; 30 minutes.

Woody Weeds	Browse Control	Fert., Est.	Fert., Broadcast	Hand-Harvest Cost	Machine-Harvest Cost	Husking/Cleaning Cost
			$23.00		$120.00	
			$10.00		$15.00	
	$10.00	$0.20	$250.00	$30.00	$20.00	
			$4.00	$5.00		$0.25
$80.00	$25.00	$0.07	$20.00	$1,000.00	$300.00	$0.50
$80.00	$35.00	$0.27	$307.00	$1,035.00	$455.00	$0.75

WOODY WEEDS. One pass on 1 acre removing woody weeds by hand. This would take 5 hours or less if it's already under control, 20 hours if not—10 hours is used as the figure here.

BROWSE CONTROL. One pass on 1 acre for egg spray; 10 gallons, presuming the coverage needed for second leaf. Less will be needed the first year, possibly more the third year. "Consumable" is eggs; this is for a backpack sprayer.

FERTILIZER, ESTABLISHMENT. A single application using the bulb planter that should suffice for two years. This is calculated per plant: 1 cup (½ pound) fertilizer and 30 person-seconds. The consumable is the fertilizer cost of $0.40 per pound (bagged).

FERTILIZER, BROADCAST. The annual application per acre, broadcast applied by tractor; 90 minutes.

HAND-HARVEST COST. The cost of hand harvest per acre for middling production. This will take less time for little bushes and more for big ones. Consumables are mostly bagging and marking supplies; this might be an overestimate.

MACHINE-HARVEST COST. This is calculated presuming middling to high production; three hours per acre.

HUSKING/CLEANING COST. Using a moderately advanced husker and cleaner, the cost per dry pound when the process outputs 20 pounds per person-hour of husked and cleaned nuts on a dry-weight basis.

The current set of numbers for good care in table 4-3 and on the top of figure 4-12 shows a little yearly net income by year 11, and it is increasing. Cumulative net income will take a few more years to reach the $0 breakeven point. First coppice and wood harvest starts around this time, too, though it's not included in these short examples; this would result in some wood income (we hope) but also some loss of nut production.

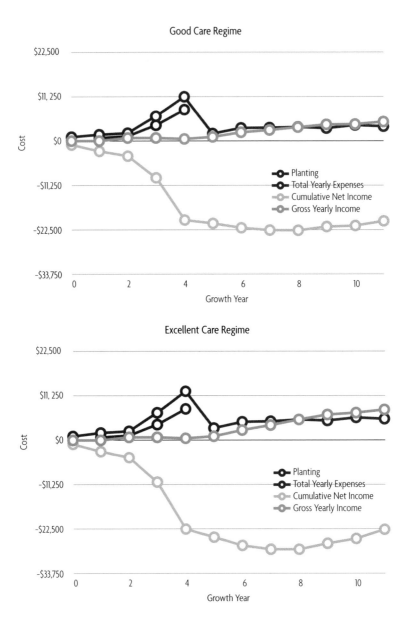

FIGURE 4-12. Income, Cost, Planting, and Cumulative Net Income for Good and Excellent Care Regimes. These graphs show that we expect good care will be substantially less expensive to begin with, but by the time maturity comes around excellent would probably give a substantially higher return.

The 10-year intercoppice cycle following this should definitely bring the cumulative net income into the positive.

We won't go into all of the details, but as can be seen in figure 4-12 we expect excellent care to cost quite a bit more to begin with; however, yearly income is also substantially greater as the establishment of the field comes to a close. This is just another example of a common phenomenon: If you can afford to put more resources into something, you'll get something back out more quickly. Not shown here though, is *more* care than we call excellent, such as mowing weekly between the bushes. Some growers do provide this level of care, but it doesn't improve the harvest potential.

As we said above, good care is what we usually recommend—the initial costs are substantially lower, and it is also closer to what most growers will be capable of doing. An additional factor to consider: Many of the increased inputs for excellent care involve extra tractor time or other carbon- and energy-intensive inputs. From this point of view, the comparison of good to excellent care in establishment is an example of *sustainable* means "slow." Good care in establishment is likely to be more sustainable than excellent, and it is slower.

That said, it may be sensible to transition from good to excellent care after the first coppice. This will take more effort and input at that time than if you had started out at the excellent level, because more replanting will be necessary and you'll be playing catch-up to bring field fertility up to snuff. If your seedling mortality wasn't too high, though, it could be possible to transition to near-excellent yield expectations by year 20. This is a slower, but probably more sustainable, way of attaining highest expected yield.

CHAPTER 5

Planting and the Establishment Period

Newly planted hazels respond very strongly to weed control, fertilizer, and reasonable water availability. The range of care you can provide runs from total neglect to luxuriously attended plantings with irrigation and individual tree shelters. All approaches can "work," but there is a direct relationship between the amount and timing of care given and survival rates, growth rate, and onset of nut bearing. (Some of this is summarized in the section *Paying for It* in chapter 4.)

One of the early breeding directions at Badgersett was to select only those plants that could survive during establishment (usually the first five to seven years of growth) even when badly neglected. This, we feel, is commonly the fate of trees—planted with the best of intentions, they often are neglected in the press of other urgencies; trees, after all, should be able to take care of themselves. Thus, all initial breeding populations underwent years of intentional severe neglect,[1] and this continues in many of our research plantings.

Results of this strategy are threefold:

1. These hazels' ability to survive terrible conditions is often amazing. A downside to this toughness is that it can take them a long time to die—giving the impression that they may survive—long after they've been damaged beyond any chance of recovery.
2. Visitors to Badgersett sometimes get the idea that the proper way to establish hazels is in heavy grass—which is not so. Some visitors also get the idea that this is what we recommend; also not so.
3. The plants with lower care may grow very slowly for several years, but what's actually happening is that they are accumulating root mass (see figure 4-3). This is a wild-type characteristic of long-lived woody plants, and our hazels have been further selected such that when they're under stress, all available resources go toward root growth. Once a critical stage is reached—and when this happens depends on fertilizer, weed control, and growing season variation—even the most neglected (surviving) hazels will take off, start to make strong top growth, and begin to bear nuts.

Never Say Probably

The single most dangerous word to utter during the establishment period of hazels is *probably*. If you catch yourself saying "That's probably enough water," or "This is probably good enough," or "This is probably soon enough"—it's not! Whatever the task is—do it until you find yourself saying "I *know* that's good enough." Otherwise—as many people have learned—you will regret it. Guaranteed.

 ## Receiving and Holding Your Plants

If you want to have live plants a week after planting, you definitely need live plants the day before planting. Once you pick up your planting stock or receive it in the mail, don't let the boxes sit in the sun, and don't let them sit in your car. Bare-root dormant hazels can be kept like any other bare-root woody plant, depending on how long you're keeping them—in the fridge can work for a week, but make sure they don't dry out! Unpack standard tubelings or other containerized plants immediately—they are *not* dormant; but *growing, and ready for immediate planting*. Once un-packed, keep them watered and *in* the sun, in conditions as close to those in the field as possible. Holding them in shade for just three days can result in de-acclimation, and heavy stress at planting. Keep them acclimated!

Around the time you receive your plants is also a good time to double-check some of the practical aspects of your field layout. Make sure your rows are easily followable (in most cases this means straight). Be sure your row ends are marked with enough room and angle for machine access, including turning. Review the *Field Layout* section in chapter 4 if you can. Fixing problems at this point is definitely annoying, but much less of a problem than *after* the plants are in the ground!

 ## Planting Day

Standard tubelings should establish well when planted from May through July in USDA Cold Hardiness Zone 5 or colder, or August for Zone 6 and warmer. We have had successful plantings as late as October here in Zone 4, but also failures. Bare-root dormant tubelings are planted like other dormant tree stock.

Survival rates over 90 percent are achieved with good practices; the average among folks who actually plant, water, and fertilize them as directed is more than 75 percent. The best have hit over 95 percent.

If you have a choice, it helps to plant as a cool, wet weather system moves in. Avoid planting in hot sun; do not start to plant until 3

TABLE 5-1. Planting Checklist for Bare-root Dormant (BRD) and Tubeling/Containerized Hazels

Task	Early Spring BRD	Late Spring BRD	Late Spring Tubeling	Summer Tubeling
Double-check lines and field tilth	X	X	X	X
Arrange crew, and task run-down	X	X	X	X
Water seedlings before planting	X	X	No water before planting	No water before planting
Spray egg			X	X
Plant	Any time of day	Any time of day	After noon if windy	After 3 PM if hot/dry/sunny
Check planting	Sufficient compaction, planting depth	Sufficient compaction, planting depth	No compaction, planting depth	No compaction, planting depth
Water in			X	X
Rake/Erosion control	X	X	X	X
Record planted locations/numbers	X	X	X	X
Mark rows	X	X	X	X
Fertilize	X	X	X	X

PM—then the baby hazels move into decreasing sun, and have a full night to grow new root hairs and start making adaptations to their new soil.

The most important factor in planting standard tubelings is to avoid crushing the roots during the transplanting process. These plants do not need a big hole—just enough to get them into the ground with sufficient loose soil to provide immediate root contact when watered in. For hand planting, we use several tools depending on the looseness of the soil: a bulb planter that cuts a plug out, a dibble spike that punches a hole exactly the size of the tube-pot (not for clay or sandy soils), and maybe a posthole digger or shovel. Standard tree-planting "bars" are deadly for tubelings: they are designed to pack soil hard around bare roots; with the tubelings, they crush the root-ball and destroy the finer roots, so this rapidly growing baby hazel has no way to get water tomorrow.

Keep exposure of roots to air and sun to an *absolute minimum*! The tubelings handle best if they are *not* watered in the tubes the day of planting, or in some cases even the day before that; soaking-wet root-balls crumble easily. To remove the plant from the tube, grasp the base of the stem just above the soil and gently pull the root-ball straight out of the tube. Occasionally a plant may not pull easily; in such a case you can blow it out with low-pressure air. Put your mouth over the holes on the bottom (using your hand to make a tube if you are squeamish about tasting a little

Tools of the Day

Here are a few tools you might want on and around the planting day.

- **GROUND MARKING PAINT.** For marking the row before planting to ensure proper row placement. Mark every plant spot if you're digging by hand; you only need to mark frequently enough to give an evenly curving row if you're using a planting machine.
- **200-FOOT FIBERGLASS MEASURING TAPES.** For putting the paint marks in the right places. We find longer tapes sometimes difficult to rewind.
- **SHOVEL.** For making holes if you're planting by hand, or for fixing errors in machine-planted plants.
- **GARDEN RAKE OR HOE.** For moving loose soil to provide proper planting depth, or to cut gully-stopping trenches on machine-planted hills.
- **WIRE STAKE FLAGS.** For marking the row after planting in case the weeds get away from you. Put them directly in the row line, but *don't* put them right next to the plant; they should be as far away from the plant as the flag is high so that they don't damage it in storm winds.
- **NOTEBOOK.** For keeping track of more than you think you need to keep track of. Write it down.
- **BULB PLANTER.** For poking a hole in the ground for fertilizer.

sphagnum) and blow hard. Hold the stem as you blow, or it can shoot out and hit your co-planters! Once it's out, handle carefully. The plants are tough, but roots and new buds are tender.

Obviously, a video or two of these processes would be helpful; we are in the production phase of the videos. Why don't they already exist? I started this project before VHS tape existed. The idea that you could easily shoot your own video, in a format easy to distribute, was out-of-reach fantasy. I've been a little slow on the uptake.

The standard tubelings are actively growing when they are planted, *not* dormant. They have small leaf areas that have been toughened by being cut back, and a small but intact and actively growing root system. Planting requires a more gentle hand than bare-root transplanting, where we are taught to stomp the trees in and pack the soil well around their roots. That will kill a tubeling, by smashing its root system so it will die of thirst. If any experienced foresters are part of your hazel-planting crew, watch them: Old habits of stomping often come back after a few minutes. Actively growing tubelings require handling extremely similar to the gentler planting long integrated into agriculture in tobacco and tomato crops. This is *not* the same as most woody plant transplantation, but it can be done by anyone willing to learn, and it can be done by machines.

Plant so the root-ball is slightly deeper than it was in the pot; ½ to 1 inch deeper is best. Covering the top of the tubeling soil is *absolutely necessary* to prevent drying out. Any exposed tubeling potting soil will

FIGURE 5-1. This is a machine-planting crew in July, in clay, in a drought. We saw survival rates of 92 percent or greater in stretches where everything was done right. We later learned that some parts weren't "done right"; survival there was less than 40 percent.

act as a wick and pull water out of the whole root-ball; exposed roots in bare-root dormant stock will stunt or kill the plant. Planting deeper than 2½ inches could hurt the plant because roots have access to seriously less oxygen at that level; some of the plants may die. If the soil you are planting into has been extensively cultivated, or "fluffed" by tilling, be aware it will settle quite a bit, and may expose the roots of the plants unless they are set deep enough to compensate for settling. That can be hard to estimate, and such plantings will need careful watching in the first months.

Machine planting is done with transplanters normally used for tobacco, strawberries, or vegetables. Not all such machines handle the tubelings adequately. Rates of more than 1,000 plants/hour are obtainable in good ground. These machines do best in well-tilled, firmed soils with few rocks; they do poorly in sods or unincorporated corn residue.

Our preferred machine planting crew consists of:

- One person driving the tractor.
- Two people riding the transplanter, one pulling from tubes and the other feeding the machine.
- Two people following the transplanter, both inspecting and fixing mistakes, one telling the transplanter crew when to stop and when adjustments are needed and the other acting as a gofer, hopefully

keeping the planter supplied with plants so it doesn't need to stop to resupply.

After transplanting, follow-up is absolutely critical. Spots where the planter furrow has not been closed properly are blaring invitations for rodent burrows, with quick damage to tubeling roots. Missing plants can be either truly missing (the transplanter feeder skipped one) or improperly dropped by the transplanter into the bottom of the furrow and buried. Buried plants can often be dug up immediately and planted properly in the right spot by hand. Walk the rows checking for holes and missed plants, adjusting plant depth as needed. If your walk-behind-the-planter people are doing their jobs, there should be few such problems. On all but the flattest fields, prevent gulley erosion by using a hoe or rake to make a small furrow across the planter furrow and wheel tracks every 10 to 20 feet along the row.

FIGURE 5-2. We prefer to machine-plant with a crew of five, but if the land is well prepared, the weather is good, and you go slowly, it can be done with fewer. This field was planted with just two, driver and planter, and one passenger—but the soil preparation was near perfect. It's a slow and gentle process.

Water the hazels well right after planting. The same day! Don't go to bed without this critical watering-in done; remember to stop the planter soon enough to leave the time you'll need. Ideally the ground around each plant should receive ½ gallon to 1 gallon. Don't dump water right on the plant; water around it. Try to water so the roots of the plant get wet, but by absorbing water from the nearby soil—this helps drive air out of the hole, and ensures good root–soil contact. Make sure the root-ball is still covered with soil after watering! (See the *Fertilizer* section below for recommendations regarding adding fertilizer to the planting water.)

Mice, squirrels, chipmunks, groundhogs, and other critters will be attracted to the nut on newly planted tubelings. If you are planting in an area where there is a lot of wildlife pressure, it may be best to gently pull or snip off the nut before or right after planting. The plant doesn't need the nut for nutrition at this point (though it will certainly use it if the nut survives). In most cases if a squirrel goes after the nut on a newly planted tubeling, it will just pull the nut off, leaving the plant unaffected; the

nut-plant attachment is specifically designed to separate. Animals
are individuals, though, and sometimes plants may be pulled out of
the ground—be on the lookout
for this, particularly in very sandy
soils, where a new plant may be
easily pulled before its roots grow
and anchor it. If in doubt, plant a
few and watch for several days to
see how they do before planting
the rest.

Protection from browsing
animals should also be done *before
you go to bed on the day of planting.* For us this means egg spray;
in summer plantings we often do
this after the sun has gone down.
The first night after planting is the

FIGURE 5-3. This diagram shows proper placement of a decapitated tubeling in the soil: ½ to 1 inch of soil should cover the top of the plug to protect against water loss and pests, and allow for settling to the appropriate level.

most dangerous night for deer damage. They know this field and what
grows in it. When they run into newly planted hazel tubelings they may
bite them purely because they were not here yesterday. Such small plants,
not yet anchored by their roots, can be just yanked out of the ground. (See
Browsing Mangement for egg-spray procedures.)

 # Care during the First Season

Now comes the hard part. The number one factor involved with success in
establishment from here through year 5 or even 10 is *owner footprint time*.
Another way to state this that I like better is an old adage: "The best fertilizer is the farmer's shadow." The point is that paying attention, getting out
into the field and looking, is the only way to notice most problems. It is
also the only way to get a good feel for what the plants are supposed to act
like, in order to then tell when action needs to be taken. It's a good idea to
learn from and through others by spending substantial time in fields at a
later stage of establishment than yours, or by asking experienced people to
come assess your field, or preferably both.

Water

For average soils in the Midwest in USDA Hardiness Zones 3b through
6b, if rain does not provide ½ an inch of water per week during the first
month, the plants should be watered. If possible, deliver the equivalent of
1 inch to each plant; in any case, check to be sure that the water you apply

penetrates deeper than the bottom of the root-ball, to avoid encouraging only shallow root growth. In "normal" weather (by Midwest corn country standards), watering should not be necessary after the first month—the roots will be settled and functioning by then. In more droughty soils or regions, watering for two months or until the plants approach dormancy is advisable. Watering in the second year has not been necessary under most conditions in the Midwest. However, in moderate or worse drought, and high heat, they do need to be checked regularly, and may require substantial watering and even extreme measures (see *Establishment in Extreme Conditions* later in this chapter). Not sure if your conditions are extreme? Check the plants.

If you have installed an irrigation system, such as a drip system or an ooze hose, it is critical to use it effectively. Water should be provided until the soil is wet in a zone extending down past the bottom of the hazels' root zone. If irrigation wets only the few inches at the surface, that is the only place where new roots will grow, and the plants will be much less effective at developing deep water-seeking roots. There's a high probability of differing infiltration rates across a field, so check the extent of water penetration in more than one place.

High heat, and winds that are especially hot and dry, can devastate a planting. Be aware that drastic measures may be needed to reduce water lost from the little plants that haven't yet established extensive root systems. (Again, see *Establishment in Extreme Conditions*.)

Unlike many vegetables that wilt easily and recover easily, woody plants seldom wilt under normal circumstances. *They can be rapidly, seriously, and extensively damaged by wilting.* If only the growing tip has wilted, they will likely recover if watered immediately. Drooping tips or leaves are signs of wilting; you can train your eye to see wilting tips—they stand out among normal seedlings.

Fertilizer

Contrary to what most of us have been taught about trees, experimentation has consistently shown that hazels benefit greatly from immediate fertilizer availability. Dr. Bert Swanson first brought this forcefully to our attention, and our later experimentation is entirely consistent with his observations. Our practice and recommendation is to provide a mild fertilizer *in the water supplied at planting (including that used in the transplanter)*. We use feed grade urea, which dissolves quickly, at a (low) rate of about 2 pounds per 100 gallons, with 3 pounds per 100 gallons of potassium chloride if possible. Potassium chloride is not compatible with all soils, however, and takes longer to dissolve than urea. It might not be appropriate for situations where tanks smaller than 200 gallons need to be mixed and applied quickly. Inappropriate fertilization is possible, but

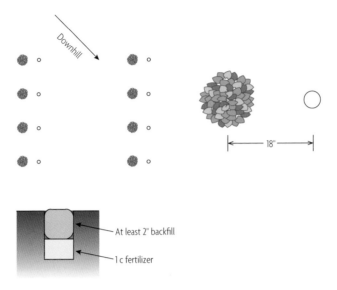

FIGURE 5-4. At the top left, two rows of young hazels are shown in green, with the small circles indicating proper placement of the bulb planter hole for the given direction of "downhill." At top right is a detail showing the 18-inch distance between the plant and the hole. On the bottom, a profile in the soil shows fertilizer at the bottom of the hole with backfill to the soil surface, which should be at least 2 inches deep.

generally mild to moderate fertilization rates will be entirely beneficial. No negative effects have been observed, survival is increased, and growth rate much improved.

Provide this fertilizer in the water at or soon after planting, or spray plants with a commercially available foliar fertilizer within the first week. In the first season of growth, the plants have not developed surface feeder roots. If you apply fertilizer by broadcasting it among the new plants, the weeds, grass, and ground cover will get much of it, and may drown out the hazels with their lush growth. Instead, you can fertilize using tree tabs or stakes, which are solid chunks of slow-release fertilizer that can be placed directly in the planting hole. They won't burn the roots, and the hazel will grow roots surrounding the chunk to feed on it as necessary. This method of application works well, although it can be expensive to apply sufficient fertilizer this way. As an alternative, we recommend the bulb planter method.

1. Punch a hole in the ground, 18 inches downhill of the plant to be fertilized, and perpendicular to the row.
2. Put a cup of 10–10–10 granulated fertilizer in the hole. The fertilizer need not be slow-release, but be sure it doesn't contain any herbicides, as "weed and feed" lawn fertilizers do.
3. Cover the fertilizer and close the hole.

FIGURE 5-5. Hazels planted two and a half years before this photograph was taken on July 14, 2011, fertilized using the bulb planter method. These are just outside the Adam Joseph Lewis Center for Environmental Studies at Oberlin College in Ohio. From the left are Griff Radulski, Sean Hayes, and John Bergen. And 15 Lace Wyandottes in the chicken tractor.

The concentrated dose of fertilizer might burn the roots of the weeds close to the hole (oh, darn), but the hazel will grow roots to find the proper distance for feeding, and will capture nearly all of that fertilizer; with no danger of harm to the hazel.

Watch your newly transplanted hazels to make sure they have sufficient fertility. Mature leaves should be dark green until they turn color in autumn. New leaves can be light green, or reddish. Pink or red young leaves are common, and do not indicate any nutrition imbalance; in fact, we think this indicates good fertility. Some young hazels normally have a red spot in the center of the leaf. For more details and figures, see *Visual Assessment of Plant Nutrient Status* in chapter 7. These guidelines apply for young hazels as well.

Weeds

Good weed control is highly beneficial. Proper weeding is essentially the same as fertilizing *and* watering your plants: It makes more of these resources available. Our weed control strategy consists primarily of cultivation and mowing. Herbicide use has proven difficult so far.

CULTIVATION. This can be problematic in wet years. In most cases at Badgersett we prefer to cultivate with an over-the-row tractor-mounted nursery cultivator once or twice in the first season (and never again), with the intention of having some mowed ground cover on that patch of the field before winter.

Do not pull or hoe weeds directly adjacent to young plants in the first two years. Pulling a big weed with its roots entwined in the hazel's root-ball can severely hurt or even kill that baby bush! If there is a huge weed right there competing for water, light, and nutrients it can be a good idea to *cut* the top of the weed down to the ground, but *don't pull it*.

MOWING. Mowing on either side of the row is effective and has several benefits. Soil is held by weed roots and mulch, which means reduced erosion. Leave the row line unmowed. Don't lose the row! The few weeds remaining in the row don't hurt. In fact, they help, by distracting deer and rabbits from the young hazels and providing a little wind protection.

FIGURE 5-6. The Mohawk of grass can protect the young hazels from deer, rabbits, and winds—but if it gets out of hand, you will see insect damage under heavy grass.

MULCHING. Mulches can be beneficial in dry conditions and for weed control, but some kinds encourage mice and steal nitrogen from the plants. Careful installation and maintenance is necessary; incorrectly done, they can kill the hazels. Mulches keep soil cool in summer and warm in winter, which may not be good for best growth and hardiness. Landscape fabric mulch has yielded mixed results. Like other mulches it can increase survival and growth rates, but requires precise installation and considerable maintenance; storm winds can rip it up if it's not very carefully anchored.

(NOT) HERBICIDE. Herbicide use is very difficult because of the high probability of damage to the seedlings; they have leaves and green bark right down to the ground. In the past, we tried using glyphosate and simazine, but simazine was found to be too difficult to control, frequently damaging the hazels. "Wick" applicators can be used to apply glyphosate without danger of drift, which is preferable to applying it as a spray. Even wick application is dangerous to your plants, however, if you hit a stem accidentally, or put herbicide on a grass stem that the wind will blow so the grass touches the seedling before the herbicide dries. We stopped using any herbicides at Badgersett over a decade ago, and this has not had any detrimental effect on establishment. We don't recommend them.

TREE SHELTERS, or tree tubes, are simply too expensive for very large plantings, and with the proper care are absolutely not necessary. We have not used them for 20 years, after a decade or so of experimenting with them. To be effective, they must be staked, weeded (yes, you have to weed the tubes), tended, and lifted in fall to allow the plants to go dormant in time for winter. They can kill bluebirds, and in wet years they can make the environment inside the tube too wet, resulting in leaf disease. The usual argument for them is "the deer eat everything!" but in fact, deer do not eat hazel leaves or twigs, outside of a little curiosity feeding.

On the other hand, several growers report that shelters definitely helped their seedlings get established, particularly the short vented versions. If you are interested in them, try a few on your site first before investing in large numbers. For young hazels 12- or 18-inch tubes are fine, particularly when they are perforated or raised for increased aeration (when mice aren't a problem, which is usually), don't have to be staked, and can be removed after a year.

GROUND COVERS. There is little research data on the effects of cover crops between rows of hazels. In the early years of a planting, avoid covers that grow tall and heavy. Dutch white clover mixed with bluegrass sown in the month after hazel planting has worked well enough in many plantings at Badgersett. The hazels will usually even survive and compete successfully with alfalfa. Velvetleaf is actually not a bad nurse crop for young hazels, as long as you don't let it get completely out of hand.

FIGURE 5-7. Mulching is tricky and often overdone. This two-and-a-half-year-old hazel has been stunted by excessive mulch depth, along with the rest of the several-acre planting it's part of. The image on the right shows it uncovered to the proper planting depth.

FIGURE 5-8. Fabric mulch is tricky and requires maintenance. This picture shows a young hazel with fabric mulch that has weeds growing on and through it; the edge was pulled up by storm winds, must be replaced, and shows the excellent mouse habitat underneath. Pocket gophers will also build mounds under fabric mulch, making them essentially invisible to predators.

Browsing Management

Deer and rabbits may attack new plantings, particularly if no other vegetation remains to distract them. Egg spray has been the most effective deterrent for deer, while raptor roosts are best for rabbits. Check chapter 6 for additional instructions on browsing management.

Egg spray to discourage deer has proven effective. Liquefy 12 eggs in a blender, mix in 5 gallons of water, and spray the mixture on the young plants until they're dripping wet. This won't wash off in rain, and is effective for two to four weeks. Do this the same day you plant, if possible, to prevent "curiosity browsing." Don't use a more concentrated mix than this; several instances have been reported in which raccoons pulled out newly planted tubelings after they were sprayed with heavy egg mixes, probably looking for an egg. If you have a lot of raccoons, don't spray egg at all until two months after planting; use an alternative commercial deer repellent if necessary.

Young hazels may sometimes be snipped off by rabbits or mice; in most cases the snipped tubeling will survive, unless snipped repeatedly. Install short tree tubes if other deterrents are ineffective or unavailable. Be on the lookout for animal damage as the seasons change; sometimes a particular mouse or rabbit will cause damage in just one small area. Weed control helps; rodents would rather not feed where they are exposed to predators. Once established, hazel bushes are rarely damaged—the critters will be a threat only to the nuts.

Year 2

Maintenance tasks in year 2 involve replanting, mowing, watering, and managing browsing animals. Timing of these tasks mostly hinges on your availability to do them, but watering and browsing management can easily become do-it-today tasks.

REPLANTING. This is often easiest in early summer in the Midwest; depending on the growing season, this can be before or after the first weed control of the season. You may not want to replant at all. If you've installed a high-density planting and mortality is only 20 percent, this can be fine. If you're doing a first-year planting nearby, then you might want to replant at lower mortality rates, since you'll be doing all the first-year chores in that field anyway. One thing to keep in mind is that the replants will require first-year care in terms of watering, weeding, and pest protection, and that's a lot more work than the similar needs of second-year plants.

FERTILIZING. If you haven't applied substantial, quickly available fertilizer yet, do this as early as you can in the spring. The basic rule is: If your hazels are hungry—feed them as soon as you can. They will use it very quickly; and we have never experienced harm to the plants by fertilizing in a wrong season. Any season is fine. (See *Target Products and Markets* in chapter 4 for a more detailed look at organic versus synthetic fertilizer.)

WEED CONTROL. Mowing is the most likely weed management practice in the second year. Don't let your little plants get completely buried in the lush spring weed growth—but do leave the strip of weeds in the row (known as Mohawking). If you've got a mower that can be adjusted high enough, one more mow over the top of the row can be useful in the first half of the growing season.

Hoeing usually results in damaged hazelnuts. Hoeing without making things worse for the hazels is a highly skilled job. Sooner or later, most folks will hoe too deep or too close and an expensive little plant that was doing fine is now dead.

It is too early to run grazing animals among your hazels. If you decide to allow grazers in your field, you need to set up fences to keep them away from the baby plants. Pastured poultry restricted to chicken tractors, however, will integrate beautifully with first year, and later, hazels.

WATERING. In dry places or years, monitor very young plants to see whether they need supplemental water. Be particularly vigilant in hot weather and windy weather or if your soils have high clay or sand content. If you see signs of wilting, water them right away! In extreme conditions, checking them once a week may not be enough.

ANIMAL PEST MANAGEMENT. Browsing animal management should be the same or similar to that done in year 1. It's a good idea to address pocket gophers as they appear, because the plants will become vulnerable to them soon (see chapter 6 for pocket gopher control information). In extreme browsing animal pressure areas, start egg spray at bud break or before, monitoring closely while new, unsprayed growth is rapidly appearing. Otherwise, monitor for damage and spray as necessary. If you are using physical animal barriers such as wire cages or tree tubes, monitor the plants for growth into the cage, and loosen as necessary—probably before you think it's necessary. If you see flags, cages, or fences coming into contact with plants during wind storms, that's a sign the barriers are too tight and are slowing plant growth.

FIGURE 5-9. The welded wire barrier around the base of this hazel is barely visible, and has been left on for at least two years too long. The constriction it has caused can be clearly seen: The middle and base of this young bush should be much wider and more dense.

 ## Years 3 and 4

Pocket gophers can be truly deadly during this time. Be vigilant about keeping them out of your field; see the directions in chapter 6.

Remove young nuts from bushes in June of the third year; bearing this early is likely to stunt their growth and reduce later crops. You can leave nuts on third-year bushes that are exceptionally vigorous (putting on more than 2 feet of new growth in year 2 or 3). If a plant isn't thriving, however, also remove nuts in year 4.

Keep track of the row! Mowing will keep your rows obvious, as will wire flags. The wire flags do need attention if they are not to become a problem later. Also keep an eye out for browse damage—rabbits and woodchucks can be locally serious at this point.

Continue mowing for weed control. If your plants are sufficiently vigorous, looking bushy and over 3 feet tall, you might be able to get a little more lax about your mowing, and even start grazing horses carefully. Start scanning for woody weeds and pulling them when you find them (mostly)—now is easier than later! Descriptions of how to do this are in chapter 7.

As you get to years 5 and 6, you may wish to protect a couple of bushes from the deer over the winter in order to provide pollen if you are in an area with no large hazels. Even though they don't browse on the twigs, deer will definitely eat all the catkins they can over the winter. If bushes are less than 5 feet tall, the deer will be able to reach all of them, and every night all winter is plenty of time to get them. Now is also the time to be working seriously on getting your harvest, storage, and marketing systems up to snuff; see chapters 8 and 10.

Establishment in Extreme Conditions

The volatility of the weather is increasing. One year may be extremely wet; another will be painfully dry. We've heard disaster stories after the fact, when most of the plants were dead, for which we could have suggested an extreme workaround that might have helped prevent so much plant loss. If you are experiencing some preposterous-feeling circumstances and your hazels are all going down the tubes, we encourage you to contact us at Badgersett. We may have some ideas for radical things to try, and we'll help if we can.

Here are some overall suggestions for dealing with extreme conditions.

FLOODING. Generally, cool, wet conditions are fine for establishing hazels, but floods can be a problem. If you are expecting imminent flooding, planting tubelings is not recommended. While mature hazels can be a good crop on floodplains, you'll need to assess your own situation and possibly find a way to plant older hazels if the flood threat is high and frequent. There are multiple ways this can be done, but our experience here is limited. Our own wetland hazels are in an old pond bottom that floods every year at snowmelt and if we have flooding rains. The hazels can be under a foot of water for 2 to 3 weeks, and are not harmed at all; but newly planted tubelings only 8 inches tall would drown.

HEAT AND DROUGHT. Extremely hot and dry conditions can be a problem for recently transplanted hazels; high wind and dry conditions can as well. The primary issue is water loss. Usually applying enough water will pull the plants through, but in record breaking hotter, drier weather growers have increasingly found themselves unable to do so—the water gets cooked out, more quickly than the roots can take it up. In these circumstances, first remove leaf mass. First, remove half of the larger active leaves. If this isn't enough, remove all the leaves; the tubelings can survive this if water balance is restored. If the soil itself is drying out too quickly, this would be the time for mulch to aid water retention. If it looks like plants are going to die anyway, then coppice them, even the tiny ones! Some will survive, and maybe none would otherwise.

INAPPROPRIATE EXPERT HELP. By now you've hopefully gotten the idea that neohybrid hazels generally, and tubelings in particular, require some things that are very different from standard orchard or forestry practice. Following advice from people who are genuine experts in those areas can be disastrous for these plants; it has happened many times. To protect against this, take the care directions from us seriously, and follow them as your first choice. We have multiple instances where we visit a customer and find a thriving, beautiful hazel planting, and the owners say apologetically, "we didn't really know anything about it, so we just followed your directions." We have many more instances where a planting is doing very badly and we hear "My uncle Larry is a forester, and has planted millions of trees; so we did it the way he told us."

When you get conflicting expert advice (including the case where *you* are the expert), set up an experiment to test whether one idea is better than the other; see the notes on experimentation in the *Field Layout* section of chapter 4. *Never* do a whole planting in conflict with the directions.

CHEMICAL RESIDUES. If you have any concern about residual herbicides in the soil where you want to plant hazels, it is best to make a small test planting before attempting a large one. In emergencies where unsuspected residues are causing problems, remove the plants immediately, wash off the contaminated soil, and replant in a hopefully uncontaminated location—but note that this is not a well-tested technique.

ANIMAL PROBLEMS. If deer, rabbits, or digging animals are creating a disaster in your hazels, adding more human presence (including camping out in the field) or a couple of farm dogs can help substantially. The next chapter in the book includes detailed suggestions for dealing with animal pests.

CHAPTER 6

Pest Management

About 15 years ago, we put out a brochure that stated blithely, "Hazels have no significant pests." At that time, it was true! Nothing truly bothersome had showed up. As the amount of hazel biomass on the farm increased, however, and as time passed, pests started to find us.

The list of diseases and pests that can affect hazels may seem daunting, but most of the time the bushes will survive and produce a crop. Significant damage generally happens only if you are ignoring and not feeding your planting. Repeatedly, we see hazels that are hungry being attacked by pests, and well-fed ones that the pests refuse to touch. As soon as you see a pest starting to get out of hand, take steps (without spraying!) to tweak the situation; that should start the pendulum swinging back in your direction.

Almost all crops have a similar number of associated pests—unless the crop was just introduced from abroad and the associated pests have yet to show up. A major reason soybeans have been such a successful and profitable crop is that in North America, they are in fact "invading aliens," brought to this country without their natural enemies, diseases, and parasites. But in 2000 Asian soybean aphid arrived, and in 2004 soybean rust. Now the cost of production can include multiple sprays, which can mean decreased production and decreased profit. In the long run, these kinds of production advantages are temporary.

If you work with an understanding of the ecosystem and allow it to balance itself, pathogens and pests are rarely economically debilitating. This concept bears some serious discussion because it is so contrary to the practices and beliefs of conventional agriculture. Standard agriculture strives to kill every single organism that is not the crop. We manage our hazels in a manner 180 degrees in the other direction. We do no (*zero*) spraying, and that includes no use of so-called benign sprays such as dormant oil or insecticidal soaps. Under this new paradigm, multiple species coexist between the rows, as well as in and on the plants. Anytime a new organism enters the system and grows into large enough numbers or biomass, something else will appear that will "eat" it. This may be a predator, a disease, a phage virus, or maybe a nematode. This management system has been independently developed by others who have also found that it works better than the chemical-heavy intervention of the old order.

Remember the Beneficials

As we were gathering photos for this book, we realized that it's been our habit to photograph mostly the pests. However, hazel fields are full of beneficial organisms, too. Look closely at figure 6-1 and you'll see one. No, it's not the chrysalis, which is a swallowtail butterfly chrysalis. The chrysalis is pretty, but hanging in a hazel, it most likely means the caterpillar that made the chrysalis ate a lot of hazel leaves. We can afford a few, but we wouldn't want a caterpillar out-break. Somewhat camouflaged on this chrysalis, though, is an insect that belongs to the family of true bugs. He's a predator—and his mouth stylus is inserted into the chrysalis, which he will drink dry. Raw nature.

FIGURE 6-1.

In his book *The Apple Grower*, for instance, Michael Phillips touches on this concept repeatedly.[1] Boiled down to its simplest form, it entails having a diversity of organisms in place, many of them occupying the spaces that pathogens and pests would exploit if they could find a toehold—and ensuring the existence of habitat for all sorts of organisms that will keep potentially damaging ones in check. It does require patience. The beneficial organisms take some time to show up following a new pest invasion of your planting. After all, if the new beneficial is going to join your farm, the first thing they need will be something to eat; your pest.

The principle that increased ecosystem complexity is valuable in limiting pest population explosions is a very sound one. We actively promote structural and species diversity within the planting. For example, one of our hazel fields has a single row of Scotch pine running through it. The trees provide a windbreak much taller than the hazels, habitat for insects that may not thrive in the hazels alone, cover and nesting sites for a variety of species of birds, and (where branches are appropriately structured) hunting roosts that hawks may use to hunt mice, 13-line ground squirrels, and blue jays. When such trees become tall enough, however, they may also provide cover for crows that steal hazels. As always, watch and manage; in this case, thin the pines. Quite a few species of small herbs and forbs coexist with hazels, and unless they cause us some specific trouble, such as burdock burrs getting into sheep fleece, we let them live there.

With this overall outlook on pest management in mind, what follows is a compendium of "organisms of interest"—how to recognize them and ways we've found to best nudge the system to favor hazel plants and their crop of nuts.

We've listed the pests you may face as much as possible in the order in which you're likely to notice them in your plantings. Be aware that many of these organisms may take 15 to 20 years to find you. Plantings less than 15 years old sometimes look pest-free, as our own did. There is no chance they will stay that way.

Diseases

Susceptibility to Eastern Filbert Blight (EFB), in combination with insufficient cold hardiness, has been a primary limitation to the development of hazelnuts as a serious crop in the northern and eastern regions of North America. The Badgersett neohybrids are now almost entirely resistant/tolerant to EFB, but there are a few additional pathogens that can and do attack hazels that growers should be aware of.

Eastern Filbert Blight (EFB) (Anisogramma anomala; Gnomoniaceae Family, Ascomycota Phylum)

The fact that *Anisogramma anomala* kills the vast majority of European hazels is the primary reason there has been no hazel industry in the eastern portion of North America. The industry in the Pacific Northwest got established because the fungus was not present there for many decades. The fungus has now found its way there, resulting in losses of hundreds of acres of susceptible plants. This has sparked an intense effort to find genetic resistance, mostly within European hazel, and use it to develop cultivars that meet market demand. The fungus is not a pathogen on the native species that co-evolved with it, *C. americana* and *C. cornuta*. It took a distribution-wide hazel-collecting trip by Philip, an ecologist with training in parasitology, to make the observations and deductions that it forms a commensal or mutualistic relationship with the native hazel species. In all the populations he and his sons visited, he found the fungus only on old or damaged branches. It appears to benefit the hazel by pruning out older, less vigorous branches and making room for the younger cohort, and preventing the plant from wasting resources. We now state that this fungus is not in fact a "pathogen" at all; but rather a fungus highly evolved to form a mutualistic relation with the two North American hazel species, but it has not co-evolved with European hazel, where its growth becomes too much for the hazel to deal with, thus killing it. Understanding that radically changes expectations of how to deal with the fungus.[2]

FIGURE 6-2. EFB forms linear cankers only on the two North American species; on the non-co-evolved European hazels, the canker will typically girdle the stem and kill it. Black pustules indicate active spore shedding.

FIGURE 6-3. Swelling at the edges of the canker indicates some resistant response from the hazel. The white pustules indicate that the EFB fungus has been attacked and taken over by one of the EFB-antagonistic microbes; several bacteria and fungi feed on EFB.

Although EFB is not a problem in plantings of Badgersett or most Badgersett-derived hybrid hazels, it merits discussion because it is still of major concern in plantings of other genetic backgrounds.

IDENTIFICATION. The fungus forms a linear series of elliptical (elongated) lesions on the stems. In tolerant plants, the length of these cankers (the collective term for the lesions) is rarely longer than 4 inches. On intolerant genotypes, the lesion can expand to an entire meter in length.[3] The centers of these lesions are black. Technically, they're perithecia (the fruiting structure of the fungus) erupting through the bark of the hazel to release spores.

In the following months the lesions that once were black with spores often turn whitish as other fungi and bacteria colonize the declining *A. anomala*.

In susceptible plants, the cankers are perennial. They increase in length and number of rows of perithecia each year and will girdle the plants above the canker. In tolerant genotypes, the canker spreads very little and does not kill

FIGURE 6-4. This orange callus is typical of an EFB-tolerant plant's response to infection.

the stem above it. Frequently an orangish callus forms that causes a slight bulging of the stem. This is diagnostic of a resistant-tolerant reaction.

LIFE CYCLE. The spores are dispersed by wind and raindrops and conceivably by the feet of birds. When transported to new tissue, the germinated spores are not thought to require an injury or even a lenticel to penetrate into the epidermis of the new host. The fungus grows in the phloem layer and outermost layer of xylem. The perithecia form 12 to 18 months following the initial entry into the stem.

MANAGEMENT. The primary means of management is genetics; selecting for acceptable tolerance levels to the fungus. Tolerance of EFB was one of the two traits selected for in the very first selection cycles we performed on Badgersett neohybrids. The success of that effort has been excellent—in fact, we state EFB tolerance to be genetically fixed within our populations, and our stock has been guaranteed since the mid-1990s not to ever die of EFB infection.

That said, this is a pest of low-vigor stems, so keeping plant vigor high further reduces any losses. Ensuring that plants are well nourished and not overmature or overcrowded are important cultural management tools.

Anthracnose

Anthracnose is the name for leaf diseases caused by many species of fungi, affecting many shade and forestry trees. It can be unsightly some years, but has not proven a consistently serious threat.

IDENTIFICATION. Anthracnose's appearance can take many forms depending upon the leaves' stage of growth when infected. Those infected when they are still unfolding can be curled and have holes in them that you may confuse with freezing damage. Leaves infected later can have necrotic (dead and crispy) margins that you can confuse with drought injury or fertilizer burn. Leaves of many ages can have irregularly shaped dead spots that look like nutrient extremes. If you've paid attention to the weather and input history, however, you can quickly rule out many of these causes. Noting the distribution of the problem can help you diagnose this disease. Anthracnose could be the culprit if most of the affected leaves are on the leeward side of the plant or on plants in a dip in the landscape.

MANAGEMENT. The Achilles' heel of this disease is that it requires water to be spread and infect its new host. Increasing airflow through the planting by orienting the rows to be parallel with the prevailing wind will reduce the period of time leaves are wet after a rain or heavy dew. Keeping current on the coppicing schedule will also prevent the plants from overgrowing the aisles, and will improve air movement. More easterly regions will have greater issues with this fungus than will the prairie and

FIGURE 6-5. To the left of the dove is a leaf that is browned and curled; this is one form anthracnose takes. It may also present as many small brownish spots, which in a bad year can enlarge and destroy the leaf.

prairie-forest transition states, where humidities tend to be lower and the wind is seemingly always available to dry things out. Using grazers to reduce foliage at lower levels in an overgrown planting can be helpful.

 ## Insect Pests

There are hundreds of insect species in a naturally managed hazel planting, the majority of which are there to feed off other insects. As mentioned before, no insect pests of hazels have reached a population size that merited chemical insecticide. A few critical pests, however, require management through cultural means or choosing hazel genetics to ensure as much high-quality crop as possible.

Stem Borer (Agrilus *spp.; Buprestidae Family*)

This is a pest that does not show up until a planting is well established—or has been seriously stressed from drought or neglect, such as serious weed competition and underfertilization. We've typically found these borers showing up in plantings only when they reach 10 years or older (another reason not to acquire your hazel seedlings from a young planting of untested genetic makeup). There is often an initial epidemic phase, before any of the natural opponents of the borers show up, and unprepared growers may be seriously worried when their consistently healthy young hazels suddenly show many dead branches. In our repeated experience, the epidemic will decline in a year or two, becoming much less significant, and allowing the resistant plants to show themselves. Typical of such relationships, the borers may have cycles of larger numbers followed by smaller numbers.

IDENTIFICATION. The wood-boring beetle larvae that attack hybrid hazels have not been identified to species. However, there are native *Agrilus* beetles associated with many tree genera: butternut agrilus, hickory agrilus, two-lined chestnut borer, and so on. Unlike the more notorious emerald ash borer (*Agrilus planipennis*), these beetle species are native to North America and share the trait of being secondary insects, preferentially attacking weakened trees and branches.[4]

The larvae are off-white with slightly darker head and no apparent legs, though the ones we've observed in southern Minnesota have transparent bristles where legs should be. There is a thickening behind the head, especially in the later instars, that is typical of the flat-headed borer group.

LIFE CYCLE. The hazel stem borer closely parallels bronze birch borer in its life cycle, and it is not impossible that they are one and the same. We are actively attempting to hatch adults in captivity, since requests for entomologists to identify insect larvae are usually met with hilarity. It overwinters in the feeding galleries inside the stem as a larva. The range of

FIGURE 6-6. Excavating a gall on this hazel stem exposes the stem borer inside.

FIGURE 6-7. The diagnostic D-shaped exit hole is where the adult borer emerged.

FIGURE 6-8. These are borer "hits" on a hazel stem that is somewhat tolerant of the insect; new vascular tissue has overgrown the damage, allowing the upper stem to live on.

larval sizes we've discovered in branches indicates there is a rather broad window for emergence dates. Their exit holes are tiny and D-shaped.

If they follow the bronze birch borer pattern, once they emerge as adults they feed briefly on leaves before mating.[5] The females then lay eggs in rough areas of the bark. The eggs hatch within a couple of weeks, and the tiny larvae chew through the bark and begin feeding on the phloem tissue. The feeding and the plant response to it form an enlarged structure probably properly termed a gall—usually associated with a node.

FIGURE 6-9. This borer larva has been mostly eaten—by an unknown something inside the borer gall—but isn't quite dead yet.

MANAGEMENT. The key is to minimize plant stress, that is, maintain vigor: Keep fertility balanced with the production level of the planting and make sure that soil pH isn't too far from the optimal range (6.5 to 7.5). Remember that coppicing can equal fertilization when it comes to rejuvenating a planting. We've found that borers are associated with EFB, with one following the other in both combinations: EFB weakens the branch, inviting borer attack, and vice versa.

We have noted at least five different ways that the plants respond to borer attack. The most dramatic is the rather abundant dead borer larvae we have found inside hazel stems. This could be due to predators, parasitoids, or elicited plant chemicals; the subject requires further study. We believe that the usual practice of pruning out the damaged stems, and their enclosed beetle larvae, is not only ill advised but most likely extremely counterproductive. You'll be removing the borers—but also everything that attacks them. In our dissections of borer-damaged stems we have frequently encountered stems where all the larvae we could find were dead, sometimes with very different appearances. Borer galls and galleries are typically inhabited by an array of other organisms: mites, isopods, and many tiny arthropods too small to identify without a microscope.

Some plants respond by growing latent buds lower on the branch; a few form large burl-like growths from repeated deposition of callus tissue. Others seem to abandon the affected stems as a natural pruning event, and the growth of unaffected stems fills the voids in the canopy. We are constantly selecting parent plants for tolerance or resistance to the borer, but maintaining good vigor is always wise.

Nut Weevil (Curculio *spp.; Curculionidae Family*)

This pest is known to most folks who have tried to harvest nuts from native hazel shrubs, because weeviled nuts are the ones the rodents leave behind for you to pick. The weevil larva chews round exit holes in the nut, and frequently has reduced the entire contents of the nut to frass (bug poop) by the time the nut would have been ripe. The nut is spoiled for human consumption, regardless. In a planting with unimproved genetics, the crop loss from weevils can be as much 50 percent,[6] but unless you're carefully monitoring the crop it's extremely difficult to quantify the percentage of loss independent of rodent theft.

IDENTIFICATION. The perpetrators are multiple members of the genus *Curculio*, beetles with specialized, oversized, long tubular mouthparts. Curculios plague many a plant group, including apples, plums, and other stone fruits. The mature larvae are cream-colored to yellowish, with dark heads and no legs. They are only about twice as long as they are wide. They are curled in the typical grub shape.

LIFE CYCLE. Adults lay eggs inside the shell of the young nutlet, leaving a tell-tale small round scar. They lay one egg per nut; the larva feeds on the kernel through the summer, then finally exits in the fall as a much larger larva to pupate in the soil near the base of the plant.[7]

If you notice an extended time period during which larvae exit the nuts, you may have more than one species involved. The exit times range from before the nut and husk are at all ripe to weeks after the nuts have

FIGURE 6-10. The telltale weevil exit holes in these hazelnuts shows the damage has already been done.

FIGURE 6-11. This excellent close-up of an acorn weevil shows one of the several species that you may find in your hazelnuts. Photo from Bruce Marlin, Wikimedia.

FIGURE 6-12. Weevil larvae newly emerged from neohybrid hazelnuts. Weevil larvae of different species are indistinguishable.

been picked and are in storage. The insects overwinter as pupae in the soil and emerge as adults in the spring one to three years later, to mate and lay eggs.

MANAGEMENT. This is another pest whose incidence and damage can be reduced by minimizing plant stress, including keeping the soil pH above 6.5. We have several times accidentally pushed our own pH below 5.0 in experimenting with fertilizer in the hazels, and in every case, weevil levels have risen dramatically, from an average of 3 to 8 percent nut infestations, depending on the individual plant, to up to 50 percent infested when the pH is below 5.5.

Whether the local pest is the North American hazelnut weevil (*C. obtusus*), the filbert weevil (*C. occidentalis*), or a hickory or acorn weevil is mostly irrelevant. At Badgersett, we select parental plants with few to no weeviled

FIGURE 6-13. The nut has walled off the weevil egg, preventing the egg from developing into a larva. This ability is both genetic and nutrition-related.

nuts even under stressful conditions; we look for multiple years' data if a plant is to qualify. We're also pleased to see some of our plants capable of walling off weevil ovipositions by laying down a woody layer inside the nutshell. This results in a nearly full-sized and viable kernel, though it may have an indentation where the wall extended into the kernel's space. So far as we know, this kind of weevil resistance has not been reported in other hazel genetics.

The use of free-ranging domestic fowl (we've had success with both chickens and guinea fowl) in the planting can result in reductions in insect numbers and an increase in well-fed birds. Fowl are quick to learn that the sound of larvae or adults plopping onto a leaf or the ground means a tasty meal is available and will compete to retrieve them. In addition, we found that guineas and chickens kept in a chicken tractor at night and allowed free range during the day caused a measured increase in pH of almost a full point in one year in the top 6 inches of soil—a substantial benefit, both directly to the plants and also to weevil management.

Big Bud Mite (Phytoptus avellanae or Cecidophyopsis vermiformis; Eriophyoidea Family)

This pest is hard to spot at first, but becomes apparent as the population builds: Flower buds enlarge, cease to develop as flowers, and die (or "blast") as a result of the mites feeding. As with the stem borer, it may take more than 10 years before you have big bud mites in your planting. In a near-monoculture of clonal hazel plantings, high mite populations can cause a serious reduction in flower buds and crop. Up to 117,000

FIGURE 6-14. These hazel buds are infested with big bud mite and are "blasted"; in pure European hazels, they will never make shoots or nuts.

Evading Mite Damage

This photo illustrates three of the five different mechanisms we've observed by which neohybrid hazels evade big bud mite. Mites are clearly present all over the twig, but the top two buds have successfully escaped infestation, and have functional female flowers showing; we've seen them bear nuts. The bud just above the thumbnail is blasted, but is showing stigmas regardless; again, such buds tend to succeed in bearing nuts. Just up the stem there is a cluster of secondary buds breaking and showing female flowers. These have started to grow because the primary female bud at the base of the catkin has been so severely affected that it will *not* function. We see repeatedly that some individual plants will signal for secondary buds to bear if primary buds are destroyed. The functioning male catkin adjacent to damaged females is also a good sign; big bud mites will sometimes infest catkins and destroy them. It's uncommon for one plant to demonstrate all these types of evasion; more often, a plant will have only one evasion technique in its arsenal.

FIGURE 6-15.

mites per bud have been counted.[8] In neohybrid hazels, at the onset of an epidemic up to 50 percent of the buds on very susceptible plants may be blasted. It is too soon to judge genetic progress in next-generation hazels where we have selected for bud mite resistance, but genetic resistance to the mites seems to be replicable and durable from one year to the next; resistant plants stay resistant.

IDENTIFICATION. The adults of this mite are extremely small (microscopic), white, and elongate, with two pairs of legs near the head. A 20× lens is necessary to see them.

LIFE CYCLE. The adults breed in the fall; damage is done during the winter when high numbers of the mites reside inside the closed dormant buds. Their feeding causes the buds to dry out and fall off or enlarge and stay attached to the twigs although dead. In the spring the females move to newly developing leaves to lay their eggs. After hatching, the nymphs develop on the leaf surface, changing into adults by summer. The colonization of new buds occurs in late summer and early fall.[9]

MANAGEMENT. We have only been able to select genetics for mite resistance since 2005, when the mites first fully established themselves at Badgersett; growers and nurseries that obtained their hazel breeding stock from us prior to that have genetics with no improved bud mite resistance. Now the extreme genetic diversity in our gene pool includes plants with multiple ways to evade mite damage; those that simply never allow the bud mite to invade, except perhaps on a few branches already damaged in some other way, so defenses are low; plants whose spring-formed buds may be susceptible, but later buds become increasingly resistant, so that there are abundant terminal buds to bear nuts and make next year's growth; and more.

Thus, if mites have become a problem in a planting, your best course of action is to update your genetics. This pest is serious enough that we vigorously advise you not to start new plantings from unimproved seed.

Japanese Beetle (Popillia japonica; Scarabaeidae Family)

This introduced pest has not yet spread to our location in Minnesota. We discovered Japanese beetles in our satellite planting in northern Illinois in August 2005 (it probably arrived somewhat earlier, but was not known in the region a few years earlier). We have been carefully watching the beetle–hazel interactions. The beetles completely consume the leaf tissue between the veins, resulting in a skeletonizing of the leaves—especially in the top portion of the plant. However, even in a row of closely related plants (such as half siblings), their occurrence—and thus their damage—tends to be extremely spotty. This is likely due to aggregation pheromones known to cause the beetles to feed in a group as opposed to spreading out over a planting.

IDENTIFICATION. The adult beetles are beautiful, with a metallic green-black head and thorax; rusty brown, longitudinally ribbed wing covers; and white vertical patches of white hairs on their black abdomen beneath their wings. Their antennae have distinctive three-pronged ends.

The Japanese beetle larva is mostly white with a light brown head and three sets of legs on its thorax (right behind its head). The hairs growing on the posterior below the anal slit form a V, which is useful for differentiating it from other soil-dwelling grubs such as June bugs and masked chafers.[10]

LIFE CYCLE. Japanese beetles spend much of their life as larvae (grubs) beneath the soil surface, where they feed on plant roots. (We don't yet know if they feed on hazel roots.) In Minnesota, the larvae pupate in June. The adults emerge in July and are active for six to eight weeks, feeding and mating. The females burrow roughly 2 inches into the soil to lay their eggs. The eggs hatch quickly, and the young larvae begin feeding voraciously; by September they are almost full grown (about 1 inch in length). When soil temperatures dip below 60°F (16°C), they begin to

FIGURE 6-16. Adult Japanese beetles coupling on hazel leaves, with considerable damage.

burrow deeper in the soil, where they overwinter as deep as 10 inches below the surface.[11]

MANAGEMENT. This is a case where we advise growers to grit their teeth and let the ecosystem come to terms with the new invader. Unless the peak of the new population crest coincides with the establishment years (1 through 3) when the plants have minimal energy reserves, no action is recommended. Controls borrowed from organic vegetable producers include use of floating row covers, securely anchored; kaolin clay sprays, which irritate and cause desiccation of the beetles; and spreading "milky spore" bacteria, which may take years to work and gets expensive fast. Because we haven't had the pleasure of their presence quite yet, we don't know if chickens or guinea fowl will search them out. The use of pheromones to attract and trap them may or may not be cost effective. The problem with these chemicals and traps is that they can attract more beetles than they can kill, so the traps should be deployed at a distance from the planting, if used at all.

Reports from multiple neohybrid hazel growers in other locations, including New York and Ohio, are that the hazels survive fine without intervention. The usual pattern is that the beetles, and damage, will concentrate on about 1 plant in 10, leaving other hazels essentially untouched. When another population surge occurs, it is usually different hazel plants targeted for beetle aggregation; no genetic factors seem to be involved in the beetles' choice.

Flea Beetle (Chrysomelidae Family)

This is another pest that took more than 15 years to show up. This means it's been present on Badgersett for a while, but it wasn't until last summer we identified it thanks to being in the field early in the morning when they're active. We have not yet identified our particular species of flea beetle, but in other crops they are known to cause economic damage. We have had a few localized infestations that did cross into the serious area, but so far damage is typically restricted to three to five adjacent bushes, becoming quickly less serious on the next bushes. We notice that on both very young (one- to three-year-old) plants and mature plants there seems to be a correlation between serious infestation and tall grass covering the leaves. They feed on the tissue between the leaf veins, resulting in what we call lace leaf appearance. The genus causing the shot-hole damage to hazels has not been identified to our knowledge.

FIGURE 6-17. Flea beetles are tiny, very evasive, and easiest to spot in very early morning. The leaf damage is distinctive. These flea beetles seem to do best in very tall grass.

The beetles are tiny (less than ⅛ inch long) and jump when startled, hence the common name. They overwinter in the soil at the base of the plants; there can be multiple generations per year.

MANAGEMENT. Reducing the height of vegetation around the hazels may help, but no other control has been deemed necessary. Cutting tall grass away can decrease the flea beetle population fairly quickly. As always, with very small hazels, be careful not to change heavily shaded plants to full sun all at once—they need time to acclimate. And don't pull grass near them; instead, cut it. We have been watching for indications of genetic differences that discourage feeding. If the problem worsens and we observe resistance, we may decide to focus our breeding efforts on increasing resistance to this pest.

Caterpillars (Lepidoptera Order)

You will see many, many species of butterfly and moth caterpillars on and around your hazels; hazels are outstanding habitat for threatened Lepidopterans. So far, due to their many predators, no caterpillars have caused enough damage to warrant management measures. This includes leaf rollers, which growers of other nuts and apples often spray for. Avoid

FIGURE 6-18. *Polyphemus* and *Cecropia* are among the huge, striking caterpillars you are likely to see in hazels every once in a while. A single one of these caterpillars can defoliate an entire branch, but there are so few of them that their impact is not significant. There are also caterpillar-like larvae; these sawfly larvae are in that category. They look very threatening when in a bunch like this, but we have not seen outbreaks of them. Do be careful about touching hairy or spiny caterpillars you don't recognize; they can cause itching, rashes, or—in the case of this Io moth caterpillar (found the hard way in our hazels by Meg) an extraordinarily painful sting. Treatment for Io moth sting: Remove the nearly invisible spines with duct tape, and use some serious painkillers.

the practice of squashing the caterpillars you can see: Odds are good that those you can catch are the ones already infested with parasites or diseases. Let them be and grit your teeth, even if you see the kind of whole-branch defoliation that a *Polyphemus* or *Cecropia* can do.

Grasshoppers (Orthoptera Order, Caelifera Suborder)

These insects can be a serious threat to small plants in the first and second year of field establishment. It is typically an early nymph stage that attacks hazel, especially if they are seriously buried in grass. Their feeding is characterized by eating everything but the midrib and veins. In heavy grasshopper years, the pests may even chew off the stem bark. They are often unevenly distributed in their occurrence and damage levels due to large broods hatching and actively feeding in a small area.

IDENTIFICATION. These insects have very large hindmost legs and antennae that are shorter than their bodies. They can jump long distances and can be carried even farther when caught by wind gusts. They

FIGURE 6-19. These first-year hazel leaves show skeletonizing by grasshopper nymphs.

go through a series of nymphal stages, during which they do not have functional wings, before becoming adults. Overwinter as eggs usually.

MANAGEMENT. We recommend leaving a bit of a weeded wind barrier in a row of newly planted hazels. This provides other things for the hoppers to eat besides hazels. But weeds can provide the habitat that grasshopper nymphs thrive in if they become too tall or thick. Removing some grass so that the young insects are exposed to foraging birds and more wind will help. Active monitoring will help you determine if a population is building. This is an instance when allowing a small group of chickens or guinea fowl to patrol the area during the day could be of huge benefit. If that is not an option for you and the hazels are sufficiently established, your best bet might be to keep removing grass until the damage stops.

Hazelnut Aphids (Corylobium *spp.*)

Although this is considered a serious pest of European hazel, we are not aware of anyone growing neohybrid hazels who has experienced aphid infestations. The aphids are known to have natural predators, so perhaps ecosystem management has done the trick. Since we never spray, we commonly find both native and Asian ladybird beetles in the hazels, as well as lacewings, assassin bugs, and many species of spiders—all good aphid eaters.

Rodents

Squirrels are the epitome of the nut-stealing pest, but gophers, mice, and their kin can be just as damaging to the yield *and to the plants themselves.* As a grower it is important to learn the players and their tricks to protect your investment from planting to nut sale.

Pocket Gopher (Geomys bursarius)

This is a very serious pest in the establishment period of your hazel planting—year 3 to year 5—and *must* be removed. They are not found everywhere, however, so check to see if you have them in your region. Pocket gophers clip off and hoard substantial amounts of root material for their offspring and for winter consumption. Prior to the third year or so, the root size is insufficient to attract their attention. By year 4, however, they become a very dangerous threat to your hazel planting. In later years when plants near maturity, these rodents can still be a nuisance because their mounds make it difficult to mow effectively and can trip up workers, but they rarely harvest enough roots from mature plants to seriously impact the plants' health and productivity. Between years 3 and 5, however, gophers can and do go right down your hazel rows, killing one plant after another, eating *all* their roots.

FIGURE 6-20. Pocket gopher and all its weapons: claws, pockets, and teeth.

IDENTIFICATION. The common name of this rodent is based on its fur-lined cheek pouch, which extends from the front of its mouth well back into its shoulders. These pouches are used to stock its larder, and it is definitely a hoarder. It has large incisors and very strong, heavily clawed front feet. The distribution of pocket gophers is disjunct, and the animal does not occur in a number of places where you might expect it. There are a number of species over North America, including the western genus *Thomomys*, which is considerably bigger than our Minnesota equivalent. They likely eat hazel roots, too.

The pocket gopher also uses its cheek pouches to transfer soil from its burrow to the surface. Because they rarely come to the surface during daylight, these mounds are often your first sign that your planting is under attack.

LIFE CYCLE. Female gophers produce just one brood of from one to six young per year unless it is a particularly good year (for them—not us!); then they may have a second. They spend the vast majority of their time in an extensive subterranean network of tunnels that you can detect by their tunnel soil mounds. A few predators dig into their tunnels and dine on them there, including badgers and canines. Some large snake species may enter the burrows and catch them. When the gophers exit from their network, they are also prey to owls (the most successful) and hawks—and the very occasional cat.

Pocket gophers, in turn, eat earthworms and grubs, but primarily the roots and shoots of a wide variety of plants. This is where the rub exists. Hazelnut roots rate right up there with alfalfa as gophers' most preferred meal. They commonly live three years and are sexually mature their second season. They live a mostly solitary lifestyle except during breeding season. Based on one data point—a trial never repeated for good reason—pocket gophers are not good to eat, in spite of being basically a large, strange squirrel.

MANAGEMENT. Pocket gophers are notoriously difficult to control. Most farmers use methods that organic farmers would prefer to stay away from. This pest, however, is not a problem that will take care of itself if left to the biodiversity of the ecosystem—at least not in human time frames, because humans have eliminated a huge portion of gophers' previous predators. An aggressive and frequently multipronged control program needs to be instituted as soon as the problem is noted, because you will be playing catch-up for some time and gophers can re-invade your area from your neighbor. Options include trapping, poisoning, or shooting the gophers; some cultural management options are effective and a help.

Several kinds of gopher trap exist. There seem to be regional differences as to which kind farm stores keep in stock. We deploy two gopher traps of the claw type linked together with a short length of chain so badgers or coyotes don't run off with a gopher and the trap. We pin the chain in place with the wire of a marker flag.

It's best to check the trapline once or twice each day. When townships offer a bounty for them, it's more common to find professional trappers for hire. Bounties differ by location, so this option might be worth checking into. The season for trapping is from the time the first mound activity is noted in the spring to about mid-October. This is when gophers head even deeper into the ground; there they live off their larders and rarely come to the surface.

Philip, our resident rodent specialist, has discovered a few tricks of the trade to improve success. In addition to inserting both traps into an opened tunnel after it makes a Y within about a foot of the mound, leave some fresh dandelion greens to mask your scent and entice the gopher forward into the clasp of the traps. You can locate the tunnel before digging by probing into the soil with a piece of rebar. Gophers do not tolerate open tunnels—which snakes, for instance, could enter. They note the air currents and return to the location to plug the entrance quickly, often within a half hour. Some rather sophisticated information about gophers' tunneling system and trapping techniques can be found online at trap makers' websites.

Placing poison pellets in gopher tunnels or directing poison gas (sulfur dioxide smoke bombs) or motor exhaust into tunnels are additional options for management. Our experience with smoke bombs is that they will kill stupid gophers, but not smart ones. At Badgersett we have dogs who help patrol the property for deer and coyotes, so ensuring they are not harmed is a priority, and use of poison very restricted. At times when an area is overrun with gophers we do sometimes use poisoned bait blocks specifically designed for pocket gophers, making sure to place them only in active tunnels and to bury them firmly to prevent predator (and dog) access.

Staking yourself out with a shotgun, sitting motionless on a chair downwind of an opened burrow, can sometimes be the fastest way to get a troublesome gopher. We prefer a 20-gauge, which will certainly get the occupant when it comes to investigate the unauthorized ventilation. Gophers are attentive, so this process usually only requires about 15 to 20 minutes of vigilance.

In terms of cultural management, keeping the area mown and/or grazed to short grass both encourages predation by owls, and makes it more difficult for the gophers to find a safer location to dump soil from tunneling. Maintaining a biodiverse environment encourages bull snakes to move in. Dogs like to dig after gophers, but their effectiveness is variable, and you need to decide if coping with the resultant holes is worth the

Predator Kites

For many of the rodents and bird pests discussed in this chapter, flying hawk- or eagle-shaped kites can reduce nut theft. Even though some of these pests are intelligent enough to know—after a while—that the thing in the sky is not a raptor, they need to divert some of their attention to it, which slows them down. A complacent rodent or smallish bird is quite soon a meal to something higher on the food chain. Kites also attract hawks and eagles that come to check them out, and these raptors provide increased surveillance of the planting.

We've found it helpful to use a larger lift kite at higher altitudes in higher and more constant wind, which helps it stay aloft. Attach the predator kite to the same string at a lower point after you've successfully launched the lift kite. On occasion, we've been able to keep a pair of kites aloft for over two days at a time. The downsides to the kite strategy are trying to launch them in erratic wind, frequently not having clear runway in the direction needed due to obstruction by hazel rows, and the inevitable string lines crisscrossing the planting when they do come to earth again. Also, the kites available for this purpose aren't very durable, providing a

nice business opportunity to someone who has a knack for kite design, since many fruit growers also use them.

FIGURE 6-21. A hawk kite flying from a pole. This allows the kite to swoop back and forth, drop out of sight, then reappear.

reduction in gopher numbers. Occasionally a cat will learn to hunt pocket gophers, and teach their kittens. If you find one, hang on to it!

13-Line Ground Squirrel (Spermophilus tridecemlineatus)

Contrary to their depiction by the University of Minnesota as robust and ferocious varmints, these ground squirrels are slender and furtive—and they are not, strictly speaking, gophers. They are diurnal (active during the day) omnivores with at least 50 percent of their diet made up of animals—grasshoppers, wireworms, caterpillars, beetles, bird eggs, and so on. The vegetative portion of their diet is made up of seeds, green shoots, and roots, which they also stash for future use.[12] They rarely climb much in the bushes, but will clean nuts from low-hanging branches and those that have dropped. They like the semi-open habitat of young plantings (approximately years 3 through 8) and are serious nut thieves.

IDENTIFICATION. These ground squirrels are black and white on a background of reddish-golden brown. Their tails are thin and are not held upright, which differentiates them from chipmunks. They are small, attaining only about 8 ounces at maturity, and are often seen standing on their back haunches to view the landscape. They are often mistaken for chipmunks, but are spotted in addition to being striped.

LIFE CYCLE. Thirteen-liners are true hibernators. They begin their winter sleep in September or early October and emerge in late March or early April.[13] Mating begins within a few weeks. A characteristic feature of their burrow that it makes an initial vertical drop of 6 inches before angling off. Squirrels usually have a single litter of from 3 to 14 young, averaging 8. Some females will have a second litter some years.

MANAGEMENT. Optimizing opportunities for 13-line ground squirrels' natural predators to nab them is effective. Keep the planting mown and/or grazed, and erect raptor roosts. Thirteen-liners are cat candy as well. We've also determined that for these and other diurnal pests, predator kites can be an effective deterrent.

Eastern Chipmunk (Tamias striatus)

Eastern chipmunks are also nut thieves—these fellows can strip bushes of nuts long before they are mature. They will sit in a convenient crotch of a bush and ingest an enormous number of nuts, leaving the husks in a pile. They are also hoarders and will store nuts for winter provisions. Until our bushes were 20 years old, we never saw a chipmunk in the hazels, although they have always been abundant in the rocky ravine only ¼ mile away.

Chipmunks are among the smallest of the squirrel clan. They have black and white stripes running from the crown of the head to the tail, but fewer than the 13-liner does. These fellows are active during the day, nesting in shallow burrows in the ground—frequently where there are rocks. Females produce one or two annual litters of three to five young.

FIGURE 6-22. Typical husk pile left by chipmunk. When your bushes are big enough to provide cover from hawks, chipmunks can move in where they have not previously been a problem. They climb the bush, cut off a cluster of nuts, process them at the base, and carry them off to store.

They live two or three years on average. In the West there are multiple additional species, of a different genus. They tend to be smaller than the eastern, but quite likely they would get into your hazels, too. They are prey to hawks, members of the weasel family, cats, and canids (dogs, foxes, coyotes, wolves).

MANAGEMENT. You have numerous options for managing chipmunks, including keeping the aisles mown or grazed, using cats and/or dogs, or setting out rat traps (remembering that hazelnuts as bait are even more attractive than peanut butter!).

Red Squirrel (Tamiasciurus hudsonicus)

These cute little fellows are not shy about scolding you for being in their territory if they can do so from a safe place. This territorial behavior is an identifying feature, as are their smaller size and coloration—reddish brown above with a white belly. When seated, their fluffy tail is most often curled up and over them.

Red squirrels are diurnal and are serious nut thieves, beginning their attack on the crop long before it is ripe. Because they are very fast, they have few serious predators. They are among a group considered tree squirrels, and typically a significant portion of their diet will come from conifer seeds.

MANAGEMENT. Given red squirrels' speed, it is even more important that you maximize the odds of their being caught by hawks by keeping aisle vegetation short. It is also important to ensure a good distance between your hazels and any other trees. This is a consideration at the time of field layout but also a maintenance issue. Prune back trees if they begin to grow into the buffer zone too aggressively. Red squirrels are quite difficult to trap.

Gray and Fox Squirrels (Sciurus spp.)

We did not have trouble with these larger tree squirrels in the first 20 years, because we started out with considerable space between hazels and woods, planting on cropland we took out of corn. These bigger squirrels would not venture out without trees for shelter. Now we *do* have them where chestnuts are thick, and they are able to reach more hazels safely. They can be very destructive, eating the hazels before they are ripe. The easiest control is to allow responsible hunters to spend the time finding them.

Unfortunately multiple growers have followed the directions of other "experts" who told them to plant their hazels right next to forest. This creates an endless struggle with the squirrels. Our rule now is to keep a 100-foot grass separation between hazel planting and forest; squirrels are reluctant to cross it, and if they start doing so habitually they will be eaten by hawks. The hawk roosts we erect help ensure that.

White-Footed Mouse, Deer Mouse (Peromyscus *spp.*)

Multiple species of this mouse genus occur all over North America; in most cases their behavior and appearance are so similar, we treat them as one pest. They will all cheerfully steal your hazels; they are among the biggest thieves both in the field and in storage facilities. They frequently chew off the end of individual nuts and eat the kernel, leaving the empty shell in its husk on the bush (or stashing it in your machinery if in a storage facility).

FIGURE 6-23. This bird nest, 5 feet up in a hazel bush, is entirely full of hazel shells, the kernels eaten by *Peromyscus.*

Both these species are semi-arboreal and nocturnal. They prefer densely vegetated sites adjacent to open fields. They are sexually mature at 8 to 10 weeks of age, their gestation period is from 21 to 28 days, and females produce three to four litters of three to five young each season. Despite this exponential production of new mice, the turnover has been

FIGURE 6-24. Deer mouse/white-footed mouse. Definitely one of the biggest nut thieves. Photo from Wikimedia.

determined to be almost complete each year.[14] They do not hibernate but enter a state of partial torpor and feed off their stored food over the winter. They frequently establish multiple caches. They eat up to 30 percent of their body weight each day, aided in their search for food by a very keen sense of smell. In addition to nuts they eat other seeds, plant shoots, beetles, bird eggs, snails, caterpillars, crickets, and flies.

MANAGEMENT. Provide raptor roosts that owls and hawks will use, and keep sight lines open by mowing or grazing the aisles. A special note here: Excluding these critters from crop storage facilities (or controlling them once inside) is important because they are hosts to the ticks that carry Lyme disease. They also host hantavirus. The latter can cause a serious respiratory disease in humans who come in contact with an infected mouse's saliva, urine, or feces. See *Transport and Storage* in chapter 8 for exclusion techniques.

Woodchucks *(Marmota monax)*

These rodents are well known for burrows that can be ankle twisters and cause equipment to bottom out or get stuck. It is less well known that they will also gnaw on young shoots in a manner similar to that of rabbits.

MANAGEMENT. Although lethal control with a .22 followed by fricassee is often preferred, trying to evict woodchucks is frequently recommended. To do so, time the eviction for late summer (early July). Before then the female may have dependent young in the den and resist relocating. Remove tall vegetation from around the burrow entrances (usually there's more than one), reducing their sense of security. Disturbing the burrow itself and using strong smells—such as by placing used kitty litter, or human wastes, in their burrow and loosely closing it—can be sufficient to cause them to move on. Dogs can also be effective tools for harassment.

 ## Birds

Only larger birds are problems in nut plantings, because hazelnuts are too large for small birds to feed on. However, birds can be a serious problem and can work stealthily. Learning the signs of their activities can alert you to their presence even if you do not see or hear them at work.

FIGURE 6-25. Note the cluster of nuts at the bottom of the branch. Crows have taken all the other clusters, very recently, from the now naked branch. Unless you observe very closely, it is easy to simply not see all the nuts—that aren't there.

Crows (Corvidae Family)

These avian nut eaters are extremely intelligent and require more than the usual due diligence to foil. They are known to be able to recognize individual humans and will post a lookout that will notify the flock of the approach of a person they know to be a threat. To thwart them, patrol your planting erratically, so they are never sure when you may be there. They react very little to being chased or hunted, but very strongly if you manage to actually kill one. The old English gamekeeper practice of hanging a dead crow where they can see it, to scare them off, actually works.

Crow pressure on your nut crop will vary from year to year, partly depending on how fast the crows learn, and partly on how big the other food sources may be that year. In a drought when corn is not available, or an off year for acorns, pressure will increase. If other food is abundant, the crows, jays, and others will be easier to scare out of the hazel fields. Pressure will also vary with the size and age of the planting—the fact that they didn't steal last year does not mean they won't show up next.

Blue Jays (Cyanocetta cristata; Corvidae Family)

The blue jay's fondness for acorns is credited with enabling the oak forests to quickly recolonize after the last ice age. Their love of hazels, with which they also co-evolved, is not as well documented. They can carry up to five acorns or hazels at a time in their esophagus and beaks. They cache large quantities of ripe nuts. Although you may not see them, learning to distinguish their various vocalizations is helpful. There are multiple jay species, all of which are nut thieves.

Having humans frequently in the field, dispersed to different parts of the planting, is one way to help protect a ripe or ripening crop of hazels. We learned the hard way not to hang tags on bushes with ripe crops too far ahead of handpicking—the jays will actually learn that the tags identify the bushes with big crops and good nuts, and they then pick the tagged bushes in the very early morning before workers can get into the dewy fields. The first time we caught them at it, we eventually chased over 40 jays out of the immediately adjacent trees.

Raptor roosts and predator kites are helpful. Blue jays do pay attention to the kite; they cannot afford to ignore it, and the startle reaction is likely to be hardwired in their brain. They still get some nuts, but the kite can slow them substantially. We fly kites in multiple ways (see the "Predator Kites" sidebar on page 126).

Of greatest importance is timeliness of harvest. Delay will result in a substantially reduced yield. Jays do not pick hazels until the husks look ripe, but they will accelerate their theft rate from then on.

Flickers (Colaptes auratus; *Picidae Family*)

These large birds of the woodpecker family are insect as well as fruit and nut eaters. They find us in sizable flocks when migrating, carrying away large quantities of nuts in their crops. Use of raptor roosts and predator-shaped kites help control populations and damage.

 ## Other Larger Mammals

Luckily for growers, larger mammals don't eat and stash away many times their body weight in nuts the way rodents and birds do, but their feeding and other behaviors can be damaging nonetheless.

White-Tailed Deer (Odocoileus virginianus)

White-tailed deer are seldom serious hazel pests, but their eating habits vary based on what else is available to them. The hazels that are most at risk are newly transplanted plants, early catkin crops, and the first plant in each row.

Whitetails are herbivores, though they've been known to be opportunistic carnivores, eating fish, birds, rodents, and bird eggs. They can cover many miles each day in search of food. Their favorite food is browse (shoots and leaves), but they also graze (eat grasses and forbs) as well as eat fruits, nuts, and fungi. In the winter in more northern locations where snow is deep, they will sometimes form "yards" where large numbers congregate. Females are capable of bearing young in their second year if their diet is adequate and the population density is low, but they usually require another year. They bear one to three fawns in mid- to late spring and soon thereafter establish their pattern of daily movements through the countryside. Most of their movement occurs at dawn and dusk.

MANAGEMENT. Applying an egg spray can reduce the eating of newly planted plants. The spray is believed to work by masking the smell of the plants, not as a repellent. Especially in high population areas, it is critical to apply a spray of a dozen eggs blended in 5 gallons of water before night falls on a newly planted field. The timing is critical, because otherwise the deer are likely to do some taste testing of these novel elements in their world. Although they don't eat much, the plants are small enough that any browsing could be lethal. This treatment should be repeated in high-pressure areas about every two weeks as new, unprotected growth emerges. Deer fences are expensive, ineffective, and in the case of neohybrid hazels not necessary. Native hazels and these deer are co-evolved, and the evidence is that European hazels are not. The hybrids appear to have enough native hazel in them to be less desirable to the deer than European hazels.

*Rabbits and Hares (*Sylvilagus *and* Lepus *spp.)*

Our direct experience with rabbits and hares among hazels is limited because of the presence of dogs on our site. Neither our experience nor the reports we've heard from growers tell us whether rabbits (cottontails) or hares (jackrabbits) are more of a pest. The diet of the hare is said to include more woody material than the rabbit's. Both are capable of producing huge numbers of offspring, with four to eight litters per year of three to eight young each. Rabbits seek refuge in underground burrows, whereas hares have aboveground nests and elude predators with speed and sharp turning ability.

MANAGEMENT. Dogs, cats, trapping, shooting, installation of raptor roosts paired with mowing and/or grazing, and the elimination of brush piles or living thickets (prime habitat) can help reduce losses to rabbits and hares. Egg spray has also been effective.

*Raccoon (*Procyon lotor*)*

Although animal matter makes up over 65 percent of a raccoon's overall diet, the seasonal shift to high-fat foods in autumn could position hazels to be up there on the most-wanted list with acorns and walnuts. However, we haven't heard of or experienced issues with raccoons except where individuals have overdone the egg-spray routine, providing, in essence, an attractive salad dressing for the young hazels. Keeping the preparation to just egg and water reduces this threat. Each species of animals is made up of individuals, however, and generalities will not always hold true. Checking for tracks and scat will help confirm a possible problem, and having dogs about the property is one of the most effective deterrents.

*Black Bear (*Ursus americanus*)*

An issue we have begun to anticipate is black bears in hazel plantings. The founder of the North American Bear Center, Dr. Lynn Rogers, told Philip he keeps a bag filled with hazels as the ultimate draw for his research bears and that he has seen instances when a bear traveled 50 miles to visit a site known to have a large hazel population.

Black bears are very rarely a serious threat to humans. Just as in blueberry territory, it may prove a good idea to generate a bit of extra noise when out working among tall ripening hazels in bear country. A 300-pound bear could also make a sizable dent in the yield of a planting if left undisturbed. Determining ways to deter or frighten away hungry bears should probably be on the radar of growers in known bear habitat. According to Lynn, though, unlike grizzlies who will bluff-charge, black bears do *not* bluff. If one seriously comes at you—do something, quick.

CHAPTER 7

Managing Mature Hazels

Once your hazel bushes have started producing nuts, the goal is to harvest as much high-quality product as possible per acre and set the planting up for many more years of the same. To do this you need to focus on four broad issues: plant vigor, plant genetics, stocking levels, and crop (nut) protection.

 ## Plant Vigor

Attaining and maintaining plant vigor is a multifaceted proposition. Just as in human health where diet, exercise, and other factors (such as stress management) are not effectively addressed singly and in isolation, the factors contributing to plant vigor are interrelated.

Mature Plant Nutrition

Now that the plants are producing a crop regularly, it is critical to ensure they are well fed and the pH of the soil is such that nutrients are available to the plants to ensure good crops every year. Keep in mind that a major component of a nut crop is protein, and nitrogen is one of the key elements in protein. Unless that nitrogen—and other elements—are added back into the system after harvest, the soil is being mined or depleted and eventually will not support a crop. The neohybrids have shown us they can keep up high nut crops for two to three years with no new fertilizer; but they will eventually collapse, both crop and plant.

Mature has more than one meaning in reference to hazels. At 6 or even 12 years of age, hazels are mature in the sense that they can produce fruit, but they are still very juvenile relative to their potential of 500 to 1,000 years of age. Badgersett has been breeding and growing hybrid hazels for more than 35 years, but that's not long enough to have seen the full scope of what these plants are capable of. We have observed that their fertilizer needs seem to drop off when they reach about 20 years old. The reasons for this change are not certain. Perhaps they become more efficient by recycling elements within each plant, or their roots have expanded to some size limit, or some other unknown phenomenon is occurring.

FIGURE 7-1. This twig burned out from lack of fertility. The kink at the end of each twiglet means there was a cluster of nuts there last year. That was a heavy load, but the nuts could have matured without harming the wood—if they had been sufficiently fed. With inadequate nutrition, the neohybrids will strip nutrients from adjacent wood to finish the nuts.

Remember also that neohybrids have been selected to bear heavier crops than wild hazels, and to bear them no matter what. If a bush does not have adequate fertility to support the nuts, they will scavenge resources from the nearby wood, killing the wood. In the worst cases plants can die back to the ground. They may revive, or not.

If you want your plants to make serious food for you, you must feed them in turn. No magic involved. The appearance of fall foliage color before the nuts should be ripe is an indicator that your plants are too hungry, and the plants are already scavenging nutrients from within to finish the nuts. Emergency foliar feeding at this stage can help next year's crop by preventing twig burnout.

Soil Testing

You need to know, for sure, the state of your field. Soil testing to guide your maintenance of a reservoir of nutrients in the soil is one way of addressing the nutritional needs of plants. You can have multiple samples from a field analyzed separately, but unless the field is very large and the discrepancies are quite apparent—sandy versus clayey, for example—multiple analyses quickly become unduly expensive and time consuming. The expense and hassle of mixing individual fertilizer batches and applying them in the appropriate areas also can be overwhelming.

I found that my local agricultural cooperative was willing to loan out their soil corer and provide bags and instructions for taking the soil samples and shipping them off for analysis. Although the agronomist at the cooperative can provide you with the raw data on your soils, don't expect them to be able to advise you on how to convert the raw data into fertilizer recommendations for your hazels. What you can do is get their recommendation for 200-bushels-per-acre corn and make the following adjustments to fit the specific needs of the hazels: Multiply the nitrogen (N) recommendation by 2 (200 percent), the phosphorus by 1.5 (150 percent), and potassium by 3 (300 percent). When you're using synthetic fertilizer, we recommend applying it with a broadcast spreader when the plants are dropping their leaves in the fall but preferably before the last rains of the season so that the rain can carry the fertilizer into the top layers of soil. Fall fertilizing does *not* hurt these hazels, and in fact appears to have the greatest impact on the following year's crop. Since the hazels don't truly go dormant, they will be actively gathering the nutrients over the winter.

It is advisable to get a complete soil test—not just the basic one—every now and then. It will tell you the status of micronutrients as well as the major players. For instance, in our region boron can sometimes be limiting. Deficiencies in these minor nutrients may seriously impact plant health and cropping.

Visual Assessment of Plant Nutrient Status

Determining the nutrient status of the soil is not necessarily the same as knowing the nutrient status of the plants. Horticulturists and agronomists have developed guidelines for foliar analysis of a great many crops, but have yet to do so for hybrid hazels. Attempts to use leaf analysis guidelines for Oregon or European hazel culture have been very unhelpful. There are, however, visual clues we've learned that can help you verify that your established plants are well nourished—or not.

- As the new growth emerges in spring, there should be some reddish coloration to the young tips.
- The leaves should be at least 3 inches long and darkish green when fully expanded.
- Distance between buds (the internode) on new growth should be at least 3 inches, and preferably closer to 6 inches.
- Well-fed neohybrids will not ever stop growing until fall leaf drop. If yours stop elongating in August, something is out of kilter—check into it. The most frequent cause is inadequate fertility.
- Fall color should occur evenly over the whole plant.

FIGURE 7-2. Look for some red color in new shoots. No red anywhere indicates low nitrogen.

The right time for this change to occur begins in early September in southern Minnesota. (Tourism websites should help you adjust for the relative date in your area.) Premature fall coloration indicates nutrient stress.

- Fall foliage color should be in the range of red-bronze-maroon. Yellow or orange indicate that the plants are hungry.

- A prolific crop of catkins going into winter is another indication of good nutrient status. A sparse catkin crop is a good reminder to beef up your fertilizer regime.

Although visual clues about plant nutritional status are helpful, it's not easy to translate these observations into a quantity of fertilizer to apply. Visual clues aren't helpful for determining which nutrients are lacking, either. A timely seat-of-the-pants application of 10–10–10 or similar premixed fertilizer is far better than nothing; you can adjust the application when you can determine the precise amounts needed. You'll need a thorough soil test for this purpose.

FIGURE 7-3. The fall color of these coppice shoots indicates that they are on the hungry side.

FIGURE 7-4. Dark red or dark bronze fall leaf colors indicate good nutrition for next year's crop.

Alternative Fertilizers: Collected Animal Manures

Organic fertilizers would fit wonderfully with our overall philosophy, but we have yet to figure a way to feasibly apply the amounts required to deliver the necessary nutrients, given that organic fertilizers are typically a much more dilute delivery system than chemical fertilizers.

Let's look at an example of trying to meet the needs of a future hazel field by applying sheep manure mixed with bedding. The soil test report recommends 100 pounds of elemental nitrogen (N), 55 pounds of phosphate (P_2O_5), and 65 pounds of potash (K_2O) per acre. In order to convert this into tonnage of manure to apply, it's important to know the actual nutrient makeup of your intended manure, how fresh it is, and how much will be available the first year.[1] Analysis of the proposed material just prior to calculation and application is important because nitrogen and potassium can be lost due to volatilization and leaching—the older the manure is, the more this has already happened in the feedlot or barn. The nutrients are bound up in organic compounds and are not all available at once—not even all the first growing season. Calculating the nutrients available the first year after application[2] indicate that 3.3 tons per acre of the manure/bedding mix will meet the needs for nitrogen and exceed the recommendations for phosphorus and potassium. This isn't too bad, as far as quantity needed per acre, but that's a lot of manure—and a lot of sheep somewhere. And many soils are even more depleted than the one used in this example, and thus would require a greater per-acre application for good results. Also, the manure will continue to release some nutrients in the two years following application, which adds much uncertainty, and could even result in toxic levels of phosphorous building up in soils. If your field does not have sufficient porosity to absorb all the nutrients, on the other hand, they may run off and contribute to algal blooms in local surface water or . . . the Gulf of Mexico.

Another wrinkle is that these calculations assume the manure will be incorporated (plowed under the surface). If this is not done, a majority of the nitrogen can volatilize (evaporate) and be lost. In an established planting of hazels, plowing is not an option. So—long story short—it may be advisable to use composted manure (less volatilization) to reduce but not eliminate synthetic fertilizer, or to combine it with other high-nitrogen materials to avoid applying excess amounts of potassium and phosphorus. The non-synthetic materials also help the health of the soil by adding organic matter and microorganisms, something the synthetic fertilizers do not. Some will find delving into this type of analysis intimidating, but others will find it fascinating and part of the challenge and complexity of practicing sustainable agriculture.

Two other ways to deliver fertilizer—especially nitrogen—to the plants are integrating grazers into the system and interplanting with legumes. At Badgersett we are trialing the use of horses and/or sheep in the aisles between the hazels. This addresses two issues—the need for fertilizer and the need for keeping the grasses and forbs short so rodents are exposed to predators to keep populations in check. Although our findings are still preliminary, we've found that horses are easily managed by timely fence change and adult sheep tend to strip the leaves; lambs can be less inclined to do so. Our data points are rather limited here: We know Icelandic sheep reliably strip the leaves but "market lambs" (Suffolks perhaps?) have been reported not to do so. This topic is ripe for further study. We believe that breed of sheep, age, and perhaps whether the lambs were exposed to adult behavior can be important variables. For adult sheep, fencing between the sheep and the hazels is probably warranted. You'll need to assess the width of your aisles, access to efficient mechanical mowers, price of fuel, and availability of your time to determine if movable electro-mesh fence is a viable alternative. When it works, it is possible to replace mowing almost entirely with grazing.

Wintering sheep in hazel plantings (no leaves present!) with supplemental hay would provide added inputs of fertility as opposed to merely recycling those present. We have tried this at Badgersett and found it a very helpful pre-coppice step: The small and dead branches were knocked down by the adults—especially the rams—and weeds and grass were entirely removed, dramatically reducing the work of coppice.

Interplanting with legumes is an attractive option for adding nitrogen to the system. Legumes fix nitrogen in the soil by forming symbiotic relationships with bacteria in nodules on their roots where atmospheric nitrogen is converted to nitrates that the plants can absorb and use. Most pasture legumes are perennials that may fix more nitrogen each year than they use—but the surplus is available to neighboring plants only as the aging nodules break down. A fair amount of that nitrogen becomes immediately available to other plants when the tops are grazed and excreted by the grazers. So the interplanting of legumes, usually in a mix with forage grasses, helps the soil fertility levels, can help nourish grazers, and also increases the biodiversity of the planting. Some growers also harvest them as hay and use them as a purse-nurse crop for income until the hazels begin to bear (see chapter 4).

Soil pH

In addition to nutrient levels, soil tests will reveal the pH of the soil. At some pH levels, nutrients are bound so tightly to the soil particles that plants cannot extract and absorb them. Keeping the soil pH at near-neutral to slightly acid levels (7.5 to 5.5) will ensure the maximum level of nutrient availability. Modifying the soil's pH is a slow and tedious process if you also have plants growing in it. Soil tends to be well buffered, meaning you must add a lot of materials such as lime or sulfur to budge its pH up or down—and it takes time for those compounds to dissolve and take effect. Monitoring soil pH periodically from the start will enable you to take corrective action before it gets seriously out of kilter and plant nutrients become locked up.

Some normal and desirable phenomena, such as the application of some nitrogen fertilizers and simple plant uptake of positively charged ions, cause soil pH to drop (become more acidic). Because of this, applying lime to the soil is a time-honored agricultural practice. "Ag lime" is pulverized limestone; calcium carbonate, $CaCO_3$.

The amount necessary to obtain your desired soil pH can be included in your soil test results. Lime can be mixed in with fertilizer and applied in the fall. However, the amount needed may be great enough that it cannot safely be applied in one application or even one year. Monitor and keep on top of this aspect of the soil that supports your crop.

Lime (calcium carbonate) is very slow to dissolve and move in the soil. Because of this, it's usually incorporated through tillage in annual agricultural systems. It can be top-dressed (and often is in apple orchards), but it can take three years for the effect to penetrate just 4 to 6 inches.[3]

Other, related compounds have a similar effect on soil pH and are useful and much more quick-acting. Hydrated lime—$Ca(OH)_2$—is a fine powder that is difficult to work with. You must use a drop spreader to apply it, or mix it with water and spread as a liquid. Quicklime (CaO) is hard to obtain and dangerous to work with (a Class 3 health hazard), but is a powerful source of OH radicals, and penetrates soil quickly. We've used it in emergencies. A pelletized version is sometimes available that's easier to handle.

Wood ashes are very basic (high pH) and can be used to increase soil pH, with the added benefits of providing potassium and micronutrients as well. Calculating how much to use is tricky, however, due to the wide variability in calcium carbonate equivalents among sources (8 to 90 percent[4]). Its fine texture and high alkalinity make it a challenge to properly store and pose a threat to the applicator and surface water if it's handled carelessly. Unless you have had your ashes tested, we recommend you don't exceed 10 to 15 pounds per 1,000 square feet.[5] If you have access to some, it's well worth using, but beware commercial or unknown sources. Where plastics have been burned, the resultant dioxins will persist in the ashes—and this is not desirable for food crop production.

We've also been exploring the use of pastured poultry to raise soil pH. The pH of fresh poultry manure ranges from 6.5 to 8.0, depending on the animals' diet.[6] In aisles where we have run chicken tractors for three summers straight, we have observed a 1.0-point increase in soil pH. Since the manure is also rich in nitrogen, the planting benefits in two ways.

Coppicing to Maintain Productivity and Vigor

Fruit growers have used pruning to rejuvenate plants for several thousand years, if not longer. Pruning removes older, less productive plant portions and increases light access as well as internal resources for the production of younger, more productive branches. In trees, unless a branch is completely shaded out it is likely to continue to grow in length and breadth for the entire life of the tree. A major difference between trees and shrubs is the tendency for shrub branches to have a finite life span, after which they are replaced in the canopy by new shoots. As crop managers, it is in our best interest to maximize the number and opportunity for the most productive branches to develop.

With our current practices, the most productive hazel stems are often around two to eight years old. On a whole-plant basis, nut production plateaus at around 10 years of age, and most of the wood is older than 6. A close inspection of individual plants will likely reveal a mass of branches emanating from the crown; among them are some that are dead or declining. The pruning technique that suits the concept of large-scale hazel production best is "prune at the base": coppicing, which is the complete periodic removal of all the aboveground portions of the plant. In addition to being a cost-effective way of rejuvenating the hazels, it also is consistent with their history. In Europe, and particularly England, there are coppices that have been maintained for multiple centuries (if not millennia). North American hazel species have evolved in ecosystems that were actively managed by indigenous peoples using fire for many hundreds of years.[7] The fires may have occurred at just about any season of the year and were capable of removing the entire aboveground biomass.

The decision of when to coppice is influenced by factors such as how much the productivity has dropped off, how much deadwood is in the canopy, and what you plan to do with the harvested wood (see chapter 11). For example, bentwood furniture requires pliable and relatively straight branches and thus would benefit from a

FIGURE 7-5. Late-winter coppice in progress. The hazel row to the left has just been coppiced, and the wood collected; the row to the right will be coppiced a year or two later.

shorter coppice cycle. On the other hand, if you are planting your hazels in successive years, perhaps you need to wait for a certain younger group to come into bearing and provide income before you coppice a particular row or section.

Coppicing is a dramatic and transformative process for the plants. In some cases—luckily a tiny minority—they simply do not regrow. In others, it's as though we put the fear of the supernatural in them and they go on to produce like never before. Still others proceed to develop an inordinately wide crown (width of the plant at the base) that is not machine harvestable without serious modification. This is why we advise growers not to decide to eliminate or propagate from a plant that has not been through coppice—it hasn't yet shown you its true colors. See figure 1-3 to understand how fast they grow back.

We recommend coppicing from February to early May, before the roots have sent winter reserves up into the canopy. We have coppiced later, and although there was less regrowth the first season relative to same-row plants coppiced earlier that year, the difference was barely apparent after three years. The plants, with a few notable exceptions, will not produce a nut crop for two years. One year, bushes bore a 50 percent crop eighteen months after coppice, but we haven't yet been able to reproduce that.

Our current coppicing practice is as follows:

1. Encircle the branches of a single plant with a rope and cinch tightly. We use a biodegradable (and burnable) jute cord.
2. While one person steadies the bundled bush, another uses a chain saw outfitted with a tungsten carbide chain to cut through all the stems as close to the soil surface as possible. The low cutting is important; you can expect to hit some dirt. If you cut too far above the soil surface, the remaining stumps become such obstacles during subsequent coppicing events that you'll have to cut at successively higher levels, and can trip your workers in the meantime.
3. The bundles are then sorted and stacked for drying and/or other uses.

Like many aspects of hazel production, coppicing is a process ripe for mechanization. There is some specialized forestry equipment that could be effectively used for this purpose. Much of it is European-made, but John Deere has also ventured into the design of forestry/biofuel harvesters. The design is typically a wheeled or track-driven unit with a boom terminated by a combination grapple (claw) and rotary blade. This would be a wonderful item for several growers to purchase cooperatively, or for an individual to invest in and use to provide contract services. It doesn't take long coppicing the low-tech way to develop a serious appreciation for what these machines could accomplish. One of the most common hindrances to greater productivity in the plantings we visit, as well as our own, are the bushes that should have been coppiced five, or more, years ago, which are disproportionately more work to deal with.

Managing Insects and Disease

Serious insect and disease pressure can be a significant stressor for crop plants, but in our hazel plantings we find that the underlying cause of most pest and disease problems is one or a combination of the big three stress factors: fertility deficiencies, soil pH extremes, and overmature plants.

When we consider the side effects of spraying pesticides, we've never encountered any instance of pests or disease that was critical enough to merit spraying (see chapter 6 for additional discussion and detailed pest management measures). Our partners in managing diseases and pests are the beneficial organisms we work at attracting and keeping resident in the planting. By planting the aisles with a diverse array of plants—or letting these grow up from the existing dormant seed bank—we're more likely to provide the habitat for beneficials.

As for damage from animal pests, we have found the plants are better at protecting themselves from herbivores than we are. Keep in mind that many plants damaged by deer in the suburbs are not from North America. Hazels are, and they co-evolved with our native animals. Given the necessary resources, they will produce anti-browsing chemicals to reduce their palatability or become somewhat toxic to macro and micro herbivores. We need to ensure they have the fertility and vigor to do so, because production of secondary compounds in response to browsing is energy-intensive.

 ## Optimizing Planting Composition

It's possible to change in-row spacing in a producing planting by modifying the density of plants within the row. Large holes with no hazels will not maximize your bottom line. To fill in these gaps, you could transplant from areas planted too densely, continue to infill using tubelings, or begin to propagate from the best of your own plants. Each of these methods has its merits and shortcomings.

Improvement by Culling

Removing inferior plants is necessary to improving the long-term quality of the planting. This is where crowd-sourcing or farmer-driven crop improvement occurs. The hybrid hazel swarm is producing an unexpectedly high number of commercially acceptable plants relative to saving the seeds from, say, apples or oranges. A large proportion (approximately 60 percent) bear an unusually strong resemblance to their seed parent.

Still, about 5 to 10 percent are below average in any number of traits, and removing them improves the population's average. The upward limit on how many to cull is up to you. The more severe the selection pressure, the better the remaining plants, but the fewer there will be and the more time and expense lost.

Do *not* use just one or two years' data to eliminate a plant—four or five years of poor performance would be a better basis. Part of what you are protecting here is the diversity of how your plants respond to seasonal weather patterns, moisture, and insect pressures. A diverse response is a buffered system. Perhaps a plant did poorly in this year's wet, cool summer, but it may be a star player in next year's extended drought. The goal is to identify and remove the plants that are subpar in a majority of the years. It requires years to compile such data. It's not necessary to develop a database system such as Badgersett's that tracks every individual. In fact, with the huge increase in the numbers of plants we oversee at Badgersett, we've also changed to a system whereby we don't enter a new plant into the database until after years of outstanding performance. We track performance by attaching weather-resistant ribbons (we use florist's tape) to the exceptional plants each year once they return to bearing after the first coppice. The color coding you choose can range from very sophisticated (perhaps using different colors to indicate superiority in different traits or for different years, allowing you to track response to different climatic extremes) to quite simple. We use the latter, with only one color of ribbon, hung on plants that possess any positive trait in the extreme. Workers in the field, whether owners, managers, or pickers, are tasked with attaching only a certain number of ribbons as they work among the plants—perhaps a number equal to 5 percent of the total; during harvest we usually issue experienced harvesters one such ribbon each day. Once three or so years of ribbons have accumulated, we put those plants with multiple ribbons into the database, and start collecting the years of data that allow us to judge with some chance of accuracy. Ideally we would also cull plants that are repeatedly in the worst 5 percent group.

Hazels Don't Die Easy

Given their size at the point when you're making culling decisions, hazels can be difficult to kill. They will be in the second decade of their lives, and they can be seriously rooted—up to 10 feet deep and 25 feet across (see *What Are Hazelnuts?* in chapter 1). Repeated coppicing, coppicing at the maximally wrong time (June), or digging with a backhoe or tree spade can all work.

Infilling with Superior Plants

Each year at Badgersett we gather additional data on plant performance to incorporate into our massive database, and consequently we continue to refine our idea of which plants are most valuable as parents for the next generation. Thus, by the time you're ready to infill an established planting, it's a foregone conclusion that better genetic stock will be available than when you started the planting. Our current recommendation for spacing in a mature planting is a 6-foot within-row spacing—this balances the benefit of a high number of plants per acre with individual plant needs for light, nutrients, water, and airflow to minimize foliar disease. The trick is correctly inserting tiny new plants into a field where the other plants are huge, cast large shadows, and have root systems extending at least 12 feet in all directions. With tubelings and other smaller plants, it may be appropriate to plant them only in large gaps or to do these plantings only in a coppice year.

Farm-Generated Materials (In-Situ Propagation)

Nursery producers can select for various traits of horticultural importance—general productivity, nut size, cold hardiness, and so forth—but individual growers can tweak the genetic makeup of their hazel fields to favor those that perform best under local conditions. You can use the plants you've repeatedly tagged for superior performance to generate seedlings or clones to improve the average performance of your whole planting.

This involves hand-collecting seeds from the superior plants, keeping them hydrated and safe from rodents, and stratifying them before planting them the following spring. It is best to wait and collect seeds only after you've culled inferior plants, so that inferior genes do not get passed on to the next generation via the pollen cloud.

Once you've removed the ripe (see chapter 8) nut clusters from the bush, they stay in their husks in a shaded area for a week or so to allow additional transfer of dry matter into the nuts. They should then be husked and kept in moist (but not soaking-wet) sphagnum at temperatures between 35 and 40°F (2 to 4°C) until midspring. If it is colder, the seed will not germinate in spring; if it is warmer, it will germinate in storage. If the seeds are allowed to dry too severely, the nuts can go into a deep dormancy and may not germinate for multiple years even though they are alive. Some growers subject the nuts to repeated leachings by soaking them in fresh water, changing the water daily and making sure it is sufficiently oxygenated. This is an added step that we have found may remove water-soluble germination inhibitors, resulting in a more uniform germination of the seed lot come spring. Because of the attractiveness of the nuts to a whole host of rodent species, it is advisable to plant the nuts in a protected area, in a covered seedbed or containers, not directly in the ground.

A second means of perpetuating your superior plants and increasing their genes in your private gene pool is by making field divisions from them. If you are in luck and they are of the type that produces small younger shoots at the periphery of the crown, you can cut off the young shoots with a sharp shovel and plant them as separate individuals. Be sure that the portion removed from the older bush already has roots associated with it.

It comes as a surprise to many who view the multi-stemmed bushes that dividing them is not an easy proposition. The great thing about field division is that it doesn't require you to have access to a greenhouse, or even a cold frame, nor to have a supply of potting media and containers. We've had over 60 percent success with direct planting of new divisions, but we've also found it is critical to provide good weed control for these until the new plant is well above probable grass growth. The timing of the process that has worked best for us is just as the shoots are expanding—early to mid-May.

In other instances we've uprooted the entire plant and subdivided it with an ax in a process we call crown division. We cut back the roots *and shoots* dramatically before replanting in their new location. This enables the roots to use the stored energy and become established before putting out leaves that will require a supply of water. This dramatic crown division process is not as time-sensitive as the peripheral division option; you may successfully accomplish it throughout much of the growing season and into the fall. Leave it too late, however, and the plants tend to frost-heave out of the soil. The trade-off for the broad window is that the aftercare needs to be very good. It's okay to leave some surrounding vegetation near the newly transplanted divisions to reduce water loss, but you'll need to repeatedly remove serious shading and deeply rooted weeds in their vicinity—preferably by cutting and not by pulling. Supplemental water and dilute fertilizer solutions of a generous amount—a couple of gallons per watering event—will encourage deep rooting and accelerate establishment.

A bonus to having replicates (clones) of the plants that perform best on your site is that you then have the opportunity to assess a plant's performance in multiple locations. This will give you a good idea of the adaptability of this plant and a much better idea as to whether the genetics are indeed superior, or if it was just a one-off fluke. It's possible that something about that particular site and genetic combination worked especially well, but without the site (say, an elk died there, or even a mammoth ages ago) the genetics are merely average. As the industry becomes established and matures, there will come a time when plantings with higher percentages of clones may be appropriate. We advise against trying to get to that destination too quickly. In part that is because once you have an entirely clonal planting, you have stopped adaptation and the buffer that a genetically diverse planting provides. Who is to say if the clone that excels this decade will continue to do so in the climate and biological challenges (say, new introduced pests) of the next? Having to bulldoze rows and acres of obsolete clones is very painful to the bottom line. Just ask growers in Oregon's Willamette Valley who planted thousands of acres of EFB-susceptible clones . . . and then faced digging out their remains when EFB was introduced to the region. As with your stock portfolio, it is smart to be diversified.

Woody and Especially Obnoxious Herbaceous Weeds

As much as we like the live-and-let-live philosophy, there are a few bad characters we cannot afford to let persist in the plantings because of the competition for light, nutrients, and space and their incompatibility with the mechanized harvest and other field operations. Woody weeds or volunteer trees and shrubs can rarely be tolerated. You may be able to welcome and even encourage these same plants in hedgerows or non-crop areas.

Controlling these weeds means being prepared to saw off, girdle, or pull interlopers whenever you encounter them, if small. For anything larger than a ½-inch diameter stem, an annual sweep is a great idea; have a flatbed trailer or a chipper in tow to remove them from the area altogether. A three-year-old box elder is easier to remove than a four-year-old, and substantially less work is required than with an eight-year-old, where resprouting is almost inevitable.

There are a couple of herbaceous weeds that are also worth the effort to eliminate ASAP. For example, garlic mustard will modify the soil chemistry and produce enough seeds that soon it produces a self-sustaining monoculture in wooded areas. Garlic mustard should certainly be controlled and attended to meticulously. Wild cucumber (*Echinocystis lobata*) reminded us just last season that some years it can seriously restrict sunlight to the plants' canopies and also

FIGURE 7-6. This mass of wild cucumber is completely blanketing the hazels beneath.

make mechanical harvest difficult (see figure 7-6). Nettles, wild parsnip, nightshade, various invasive thistles, and burdock can all be problems if you let them get out of hand; whatever qualifies as a noxious weed in your area is likely to warrant attention. On the other hand, nettles and some other "weeds" may be quite profitably harvested if you have the workers required.

Wild parsnip (*Pastinaca sativa*), newly invasive for us, is treacherous due to the poison ivy–like phytodermatitis its sap causes in sunlight. Its ability to regrow after being mowed means you have to dig out individual plants, unless you mow repeatedly during flowering, or maybe graze the sheep on them, depending on your predilection.

Protecting the Maturing Crop

Having a vigorous, well-stocked planting will get you most of the way to a bountiful harvest. The final step is making sure that you (and not the multitude of non-human nut lovers) are the one to harvest the crop. Protecting the crop starts long before the first nuts are ripe. Chipmunks and squirrels will start to harvest nuts as early as midsummer. Too many of these little thieves—and those that come later—will seriously erode your profit margin.

The installation of raptor roosts (see chapter 4) is a great start, but improving visibility by removing tall vegetation in the aisles is important in order for the raptors to see the rodents and perform their swooping population control. You can remove vegetation by mowing or turning grazing animals into the area. This needs to be done at least once prior to the final clearing just prior to harvest. It is by far the best to mow or graze twice or more per season. The raptors need a place to eat in early July just as much as they do in mid-August, and if you don't provide it they'll go elsewhere. One year our first mowing was just before harvest, and it took the diurnal hawks at least two weeks to rediscover our *Spermophilus* smorgasbord.

All mowing is not equal. A flail mower will cut the stems near ground level and pulverize the cut material, whereas a sickle bar will merely fell it over, which provides a nice covered pathway for the rodents to exploit. Sickle-bar-type mowing can also create a dry straw thatch that can become a fire hazard.

As I mentioned in the fertilizer section, grazers, especially horses and sheep, seem to be a natural pairing with woody ag crops. We are still experimenting with this, but the benefits so far seem to be that they can graze down the vegetation within the row as well as in the aisles, giving superior sightlines to the raptors. They can also be substantial allies in the control of some less desirable companion plants such as thistles, nettles,

FIGURE 7-7. Grazers among the hazels removing cover from the rodents and cycling nutrients.

and parsnips. For them to do this best, it helps to give them access to the planting from early in the season when the weeds are palatable and bring them back through when regrowth has occurred. Livestock can be quickly rotated through and still do a substantial job of vegetation control. It is not a possibility to use them on rows that have just undergone coppicing, however.

In addition to keeping the view open for raptors, continuing to trap pocket gophers (see chapter 6) is a good idea—especially if there are smaller tubelings or divisions interspersed in the planting that could be killed off by a gopher strike. Running over and eating gopher mounds can overstress machinery, driving over gopher mounds is hard on your back, and the tunneling and mounds can tip over and ground out electromesh sheep fencing.

By the time your planting is producing nuts, flocks of large, nut-eating birds will have found it. Because they can work the fields quite silently, you may not be aware of how many there are and what an impact they have on your harvest.

Means of reducing their take include installation of numerous raptor roosts, flying kites (see chapter 6), and actively hunting them, especially crows. Doing nothing is really not an option. They will eat all your profits, if not this year, then next.

CHAPTER 8

Harvest

After some years, and quite possibly some tears, a spring will arrive when the deer have left enough male catkins uneaten to allow hazel pollen to pollinate the tiny female flowers. When this happens, you'll start to see clusters of nuts developing! Most hazels will show substantial development of fertile clusters by mid- to late June (in Zone 4), but some may be hard to see well into July, even for the trained eye.

This is an exciting time! Harness that excitement and keep a close watch on your nuts as they develop, weekly throughout the summer and every couple of days as ripeness nears. Unless you have a clear idea of how many clusters have been developing, you may find it difficult to discern whether nuts are disappearing, not to mention how. If your hazels are in the middle of miles of tilled ground, nut thieving may not be a big problem for years, or even decades. But if your planting is near a population of deer or squirrels or mice or blue jays (including those in town), watch out! Like as not, they'll get the nuts before you do the first year.

Before you do harvest, though, it'll be handy to have a few things ready to go. Here is a list of items we recommend for harvesting:

- Picking bags, of the sort used in apple orchards, or trugs (10-gallon bins into which nuts can easily be tossed, if you will be picking by hand). Five-gallon buckets can be used, but take more effort.
- Harvest transport bags: white plastic trash bags if you're picking early to avoid loss to pests, mesh (preferably raschel knit) if you're picking dry.
- Long pants and long-sleeved shirts (old ones; they'll get stained).
- Hand nut cracker(s) for double-checking ripeness, fill, and so on.
- Gloves, for handling weeds and/or keeping your hands from being eaten by the husk acid.
- Eye protection is a good idea. Stick in the eye is not fun.
- Tools, to cut or dig woody weeds, thistles and nettles. At least a handheld pull saw and a sickle, possibly also a sharp shovel, lopping shears, and an ax. Chain saw if your hazels are badly overgrown.

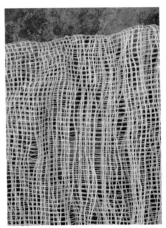 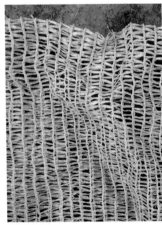 

FIGURE 8-1. Mesh bag types commonly available in 2014. The square-weave bag (left) is the standard for onion or fruit bags designed for tens of pounds of product, although the weave will deform and let loose nuts fall through. The raschel knit mesh (middle) is available in similar sizes and is less likely to let free nuts through. The Vexar mesh produce bag (right) can hold several pounds of husked nuts but is usually neither strong nor large enough for field use.

- Florist's tape for marking outstanding bushes. Give each member of the picking crew one to three markers per day. (See *Improvement by Culling* in chapter 7 for more on marking bushes.)
- Camera. You *will* see something interesting; take photos for your records and to share.
- Wire stake flags, to mark holes/hazards, such as wasp nests.

When to Harvest

Wild hazelnut shells usually turn a tan to dark brown when ripe. At the base of the nut where it attaches to the plant is the hilum: the light-colored structure across which nutrients are pumped from the parent plant into the nut. Nuts that are ripe will be at various stages of separating from the husk at the hilum; unripe nuts will tear pieces of the husk along with them if they are removed from the cluster. Neohybrids may be white, or even leaf green, at ripeness; this can cause some confusion for handpickers, but also very possibly for birds and mice, which stick to ripe nuts.

The shell has two halves and a smooth interior, giving rise to a roughly spherical or ellipsoid kernel. The kernel itself is covered with a thin, sometimes fibrous brown layer called the pellicle, and has two halves as well. When freshly ripe, it will often fill the shell nearly completely. Kernels shrink due to water loss during drying to a degree that varies widely with genetics.

Starting in late July, scan plantings for ripeness. *Do not* wait until you see brown, "ripe" husks. Almost all new hazel growers carefully watch their first crop ripen, and keep watching . . . until the nuts are all gone.

FIGURE 8-2. All the nuts from one bush look nearly identical. Each of these nuts is from a different bush, and illustrates the great genetic diversity we have. The nuts are also this variable for the traits you do not see such as taste, oil content, shell thickness, and so on.

Hazelnuts are ripe inside their husks as much as *two weeks before the husk shows ripening*. It is during this window that you can pick by hand and outsmart (most of) the critters.

In order to beat the critters to the party, you need to know the secret ninja way to check hazelnut ripeness. Push the top of the nut sideways, hard. If the hilum separates and the nut pops free of the husk, it is ripe enough to be picked. This separation is called abscission, and is shown in figure 8-3. You'll encounter some variability, but in most cases the husk will show little or no ripening/browning when the stage of abscission is

FIGURE 8-3. The nut on the left is not ripe enough to pick; the hilum has not abscised and is still attached tightly. The nut on the right is just ripe enough to pick; the hilum has abscised sufficiently to allow the nut to come loose from the husk.

FIGURE 8-4. Green-ripe (top), brown-ripe (center), and dry-ripe (bottom) clusters.

reached. If left on the bush, the nut will then usually progress through "brown ripe" to "dry ripe," as shown in figure 8-4.

Ripening dates in southeast Minnesota (USDA Zone 4b) and eastern Nebraska (Zone 5b) differ very little, but can be a week or two later than northwestern Illinois (Zone 5a). The hardiness zone alone is not a good predictor of ripening date; you just need to keep track of how ripeness progresses in your area. The variations in the current year's growing season will affect ripening dates. Nuts often ripen early in a very warm season, and may be early or late in a cool season. Drought may tend to delay ripeness; in extremes it may also reduce the percentage of the crop brought to full ripeness.

Non-humans don't check nut abscission to judge ripeness, but there are many other hints, some of which will signal nut ripeness to the birds and beasts as well. On many bushes, the fleshy portion of the husk surrounding the nut will turn a slightly rusty brown before the husks actually start drying. This husk coloring is a strong visual signal to many nut eaters, and by the time you see it you may be losing a lot of nuts. The brown color of the nut itself is also a hint for humans, but nut color is by no means a full indicator; we have bushes whose nuts are white, green, or mottled when ripe, though they all turn brown eventually (so far). Leaf color is another hint; fall color often won't show strongly until the crop is well ripe.

In most cases about 80 percent of the nuts on a bush will be abscising and ready for hand harvest at the same time. Pick all the nuts on the bush when this happens, if you're harvesting by hand—wait for the last 20 percent and you could lose them all.

When judging ripeness, keep in mind that some nuts are more likely to be among the final 20 percent. Examples are bottom nuts in a big cluster, nuts in shady places, and nuts on damaged branches. Be careful, though: Nuts on damaged branches can ripen early instead. To determine the ripeness of a single bush, check at least three clusters on the bush, and not all on the same side.

If you're managing a big crew of handpickers, you may have to judge ripeness and pick row by row rather than bush by bush; try the same ratio of 80/20 ready/unready for handpicking, but over the whole row rather than a bush. Judging ripeness of an entire row may require 10 or more samples (from five or more bushes) on each side of the row.

Judging ripeness for machine harvest is a bit different; see the section later in this chapter.

Who's Stealing My Nuts?

No matter how carefully you monitor ripeness, it's pretty likely that *some* sort of nut theft is already going on. Here are some pointers to help you figure out where they're going.

PILES OF NUTS OR CLUSTERS NEAR THE BASE OF A BUSH. Squirrels or chipmunks if shells are broken in half, mice if they've got holes chewed in them.

SINGLE NUTS MISSING FROM CLUSTERS. Mice, blue jays, or woodpeckers. The birds will often leave a peck mark in the husk, with part of the husk broken off. If nuts are missing high on the bush, the culprits are more likely birds, but mice climb way up there, too.

ENTIRE CLUSTERS MISSING. If the clusters were torn cleanly off, it's probably crows. If it looks like the husk stem is cut rather than torn, it's probably squirrels (but these are more likely to escape your notice).

MANGLED HUSKS WITH OR WITHOUT NUTS MISSING, 3 TO 5 FEET ABOVE THE GROUND. Human checking for ripeness.

CRACKED NUTS AND HUSKS SPREAD OUT AROUND OR NEAR A BUSH. Deer (particularly if the shells are thin).

How do you slow down this nut theft? For most pests, the best defenses are raptor kites, human presence, mowing, restrictions on pest habitat highways, and possibly egg spray (which we have tried several times and will try again). For more targeted anti-theft measures, see chapter 6.

How to Harvest

Picking ripe nuts off the bush is a simple enough process, but as with any harvest, there are details that make the difference between fun harvest and drudgery, or profitable harvest and throwing resources into a pit. Harvest can be done by hand or machine, and the concerns for each are quite different.

Hand Harvest

Handpicking is a learned skill, and takes effort. Managing pickers is also a skill, and takes effort. Wear old clothes—they will get stained (by bugs, spiders, and the bush itself).

Be aware of the bushes, branches, and twigs that will bear next year's crop. Pushing carelessly between bushes, or through them to get at nuts in the middle, can break branches. Pulling off a bit of the twig that each cluster stem is attached to can take next year's flower buds (and crop) with it as well. Some bushes can rebound by quickly recruiting secondary buds, but you should still try to avoid picking off those buds. Happily, you avoid this in the same way you make harvest less work—by giving the cluster a little twist as you pull, or by pushing the stem of the husk to the side as you pull. These motions will feel somewhat familiar to experienced apple pickers. Sadly, the precise motion to best pick off clusters varies from bush to bush, so it takes some extra attention to adjust as you go down the row. Happily, with experience this adjustment can become automatic.

Put the nuts in your picking bag or toss them into the bucket, trug, or fruit picking shoulder bag as you get them off of the bush. For taller bushes, it can be useful to have one person pull down a branch while another gets the nuts off of it. Once your picking container is about three-quarters full, dump it into a harvest bag (either white plastic or mesh, depending on when/what you're picking).

It's not easy to spot all the nuts on a bush. As you move around a bush or down the row while picking, look behind yourself as well as in front. Some nuts will only make themselves known when viewed from a particular vantage point. When you see a cluster on the other side of a bush, pick it as soon as you spot it if possible; it may be impossible to find from the other side. The nuts do hide—experienced pickers find many nuts by first noticing the slow sway of a branch, which indicates the weight of a nut mass. Or they feel along each branch, checking for clusters at the nodes. This method is often faster than a visual search. If you try feeling for nuts, keep in mind that there can be stinging things in the bushes, too.

It is pretty much impossible to pick the last nut on a bush; "there aren't any more" is a common harvester's joke. Because of this, the picking crew chief will need to say "It's done!" to move the team on to the next bush. For most bulk harvest, if a picker is looking for the next nut for more than three seconds it's about time to move on. For experienced harvest managers, it can also be useful to set goals for the crew: "We should be able to finish this row in two hours," for instance. Making these estimates does take experience, but it's often possible to go faster than you might guess. People used to seeking out every last nut no matter how long it takes will need both goal setting and psyching up to reach their full potential. If you're doing hand harvest on a bush-by-bush basis, having two people each go around the bush completely, in opposite directions, can yield several percent more than having just one do it. If you're doing bulk hand harvest of a row of bushes, each person can go down one side of the row at a time, not attempting to get all the way around each bush, but not forgetting the nuts in the middle, either.

Machine Harvest

Given the goal to build a hazelnut industry, mechanized harvest is an important piece of that picture. Some picking machines take nuts or fruit from bushes and trees directly; others work by shaking the nuts or fruit to the ground and then sweeping them up. The direct-from-the-bush method is the pathway of choice for neohybrid hazels, for multiple reasons. It avoids a slew of problems associated with contamination on the ground, requires less weed removal and ground leveling around the bushes, and is likely to also allow earlier harvest with fewer losses to animals.

To date, the blueberry-picking machines made by BEI, Oxbo Korvan, and Littau look like the best bet. These use various mechanisms to shake the crop off the bush, then catch it on pans underneath that, direct it to conveyors, and run it past blowers to remove leaves before depositing it in lugs or bags. Since 2011 Badgersett has been harvesting with a small self-propelled 1960s-era blueberry picker made by BEI of Holland, Michigan, that we helped acquire for one of our larger growers. It uses a beater-style bush shaker; there are also swayer, vibration, and even fan/blower bush shakers available. The 6-foot-tall tunnel of this harvester is too short for some of our rows that have gone too long without coppice. The piece of the mechanized management picture we need is either a coppicing machine or a bigger harvester—there are many bigger harvesters nowadays, but road transport becomes an issue for most of them. With the current set of small fields widely scattered across the Midwest, being able to legally haul the 8-foot-wide harvester on a flatbed trailer without worrying about wide-load permits is a real plus.

This blueberry harvester basically works for hazels! For the past three years we have harvested more by machine than by hand, and we expect that to continue. There are many possible improvements waiting to happen. Conveyer modifications in general to deal with the higher volume

FIGURE 8-5. "Blue Boy." This 1960s-era BEI blueberry picker was converted into the first hybrid bush hazel harvester.

of harvest trash from the hazels (compared with blueberries) would be useful; modifications to allow flow into high-capacity trailed or forklift bins would also be good. If you're thinking of buying a machine, beware the fancy conveyor buckets on some of the newer blueberry pickers. We suspect that these buckets aren't tall enough for hazel clusters to fit in, or that the buckets will spit out the clusters at the wrong place in the machine.

These styles of picker are quite gentle with the bush and the harvested crop, even though they look like they're beating the bejeezus out of them. On the upside this means that the bushes, next year's crop, and small animals that pass through the picker are all generally unharmed. On the downside, many of those small animals are insect pests. Due to this, and the additional simultaneous harvest of weed seed (wild parsnip, multi-flora rose, wild cucumber, et cetera), it is not sensible to transport the crop in their husks over long distances. They need to be husked before such transport or you will spread pests, which can be an offense with a large fine attached.

So far making two harvest passes per season seems to be the most effective with our current machine and current genetics. The first pass should be when you can expect about 50 percent of the nuts to come off of the bush. Note that this is much later than when 50 percent of the bushes are ready for hand harvest. Using the machine earlier results in 10 to 40 percent of the picked nuts being unripe and failing to come loose from the husk. The crop therefore needs to be cured before husking, as in hand harvest (see *Transport and Storage* for curing instructions). Picking

FIGURE 8-6. Praying mantises, spring peepers, baby goldfinches, and even eastern red bats may show up among the nuts in our mechanical harvester.

the nuts early appears to result in approximately 10 percent lower final dry kernel percentage in the nut. Waiting until about half of the crop will come off the bush (and nearly that much is dry-ripe) usually results in the clusters being huskable immediately following harvest.

How do you test expected harvest percentage? Beat the bushes with a stick, carefully. You can come up with a reasonable estimate by taking a stick about the same size as the harvester beaters and moving it in a manner that approximates the motion of those beaters as closely as possible. Be advised that people may think this looks very funny, even if they know what you're doing and why. Also, predicting harvester performance this way can be tricky; the percentage of nuts that come off can be quite sensitive to changes in force, speed, and angle of beating, and may vary from day to day. Beating the bush is preferable to shaking the bush with your hands, because shaking creates a very different motion than the beaters would, and does not give an accurate simulation of the harvester. Shaking with your hands usually will remove more nut clusters than the harvester will.

The second harvest pass should be when 99 percent of the bushes are ready for handpicking; most of the field will be dry-ripe at this point. Aim to leave 10 to 20 percent of the nuts on the bushes following this pass; if you wait to harvest until all of them will come off, you'll most likely lose more than that 20 percent to pests, or to incidental loss as nuts become very ripe and drop off the bushes naturally.

We would like to provide a summary of data regarding harvest times, efforts, and efficiency for this harvester, but the reality is that we don't have enough data yet. Here's what we do know:

- Using a crew of five (including the driver) results in the least machine-idle time, but three works pretty well for most situations. Some modifications allowed Brandon to run the harvester with no tending crew on some sparser rows last fall, but it did still result in a higher percent of field loss. In general, lower crew numbers result in more frequent machine stops and higher field losses due to clogged or misadjusted equipment; where the breakeven point is depends on how much you're paying the crew.
- Near-maximum rates of harvest so far (not including passing quickly over a lot of mostly empty bushes): 1.5 acres per hour for smaller bushes, where the ground speed required to beat each bush sufficiently is the limiting factor; twenty-five mesh harvest bags per hour (in as little as 0.6 acre) on larger, more appropriately sized bushes. Those bags weigh 30 to 40 pounds at harvest, but yield 5–8 pounds or so each of dry, graded, in-shell nuts.
- Machine harvesting yields more nuts per harvest bag than hand-picking, because the beater mechanism will dislodge many nuts from the husks as it picks, leaving more nuts and fewer husks in the machine output.

All these harvest rates and crew numbers will be changing quickly as equipment and practices develop over the next few years.

As you are harvesting (and indeed spending any time in the field), be aware of possible dangers and pitfalls. One is stepping in a fox or woodchuck hole and twisting an ankle. Mowing the field just prior to harvest reduces the risk of this by exposing previously hidden holes; it also improves sight lines for raptors. Other potential dangers in the field are being spooked by spiders, stung by wasps, or pierced with the sharp hairs of certain caterpillars. As you enter the Zen of handpicking —especially as you graduate to picking-by-feel—you may not see the threat until it's too late. Dressing in long sleeves, tucking pant legs into socks, and being alert are all helpful. Question prospective pickers as to whether they experience anaphylaxis from insect stings; if so, both they and the crew leader should carry the antidotal EpiPens at all times. It's a jungle out there!

Transport and Storage

The nuts are living things, even if dry. They need to breathe both during harvest and post-harvest. With few exceptions, live nuts will not spoil whereas dead nuts quickly give in to mold and rancidity.

When transporting fresh nuts, either with or without the husks, pay special attention to allowing ventilation and avoiding overheating. Even dry-ripe nuts are likely to have enough moisture straight from the field to cause condensation inside a closed vehicle, and we have experienced cryptic spoilage that was hard to explain other than by "it wasn't ventilated well enough in transport right after harvest." The combination of heat and moisture with insufficient ventilation and lack of cooling appears particularly problematic.

If you are picking by hand based on abscission date rather than husk ripeness, you will want to let the husks cure for a time before husking. The most effective method we've found is picking into 20- to 30-gallon white plastic bags and leaving the bagged nuts sitting in a protected, shaded spot for one week. The increased humidity inside the bags allows considerably more ripening to happen than if they were dried down immediately. Starting immediately after picking, keep the unhusked hazels in *closed* bags (with drawstrings pulled shut, leaving a small air passage) in the *shade*. Don't leave the closed bags in the sun, not even for a few hours in the field; the nuts can overheat and suffocate there. By the end of the day, transport the bagged nuts to a mouse- and animal-proof building, and put them where they will be shaded for the next week.

After seven days, open the bags and transfer the nuts to drying conditions; for nuts harvested dry-ripe, you can start here. Leaving them in closed bags longer than seven days can cause suffocation and spoilage of the nuts.

The drying area needs to be mouse- and raccoon-proof, too, as well as ventiliated to allow air through. Keep the nuts in mesh bags, or unbagged, in the sun. Unbagged can be better for large quantities if you've got the equipment to turn the nuts over and move them in bulk. Turn the nuts daily under these conditions to avoid mold developing among the less-well-ventilated nuts on the bottom. Don't dry them down too hard before husking; most husking machines work better with a little moisture still in the husk!

Mouse-Proof Storage

Hazelnuts are estimated to be at least 10,000 times more attractive to mice and other grain eaters than a grain of corn. (This is based on Philip's longtime study of sensory perception in rodents and actual measures of the characteristics of filbertone, the chief component of standard hazel

FIGURE 8-7. A building fit to keep nuts in. The floor is concrete slab, the walls are ¼-inch galvanized hardware cloth, and the roof is isolated; the unclimbable strip of siding at the top of the exterior wall prevents mice from reaching the roof. Note, however, that this strip is actually failing at the moment, due to the stick left dangling across it right in the middle of the picture, which forms a highway the mice will appreciate.

flavor.) Protecting a huge harvest of nuts from rodents is a serious challenge. Having the building envelope—walls, roof, and floor—built of something mice cannot chew through, and no joint crevices greater than ⅛ inch, is a start.

If you are constructing a nut storage building or bin, it is necessary to keep in mind all the things that the storage structure needs to keep out. Juvenile deer mice are the smallest creatures we know of that regularly chew through hazel shells, but the substantially smaller gray shrew and only slightly smaller short-tailed shrew will both gather and cache hazelnuts when they can find them. The walls, doors, floors, and roof of the storage area need to keep out these—as well as raccoons, crows, and possibly even bears—in your area.

CHAPTER 9

Processing:
From Harvest to Market

Once you've gotten your nuts out of the field, you'll have to dry, husk, grade, and clean them in order to sell or use them. As with grain crops, even in the bulk commodity market, prices are much better for a husked and dried product. In this chapter, we'll take you from husking to cleaning, sizing, and grading husked in-shell nuts, quality assurance and flavor development, cracking, cleaning kernels, and the distinctly value-added and food-grade step of roasting. All of these steps can be done by hand, but once your harvest brings in a couple hundred pounds of nuts, equipment of some sort becomes a must. Since the existing hazel industries rely primarily on plants with smaller husks, fewer nuts in a cluster, and clonal plantings with uniform nut size and shape, most of their machines do not readily adapt to use with neohybrid hazels. They are a new crop, and we are just beginning to have enough crop available to test, develop, and refine higher-capacity machines. In this chapter we share our extensive knowledge about processing neohybrid hazels on a small scale, and informed hypotheses about working with them on the large scale.

 Husking

Husking is the removal of the husk, or involucre, from the nut. It can be done by hand, but this gets old pretty fast. Usually in about 45 minutes. A step up from husking by hand is doing it by foot—either by having the kids jump up and down on a sack full of nuts, as grandma used to do it, carefully stepping on them in a container of some sort (many feed tubs are tough enough), or doing the twist on a platform in said tub. Beyond 100 pounds of nuts a year, though (and sometimes before that), you're going to want a husking machine. There are numerous designs for husking and husking-cleaning machines made for neohybrid hazels, and even a couple of machines you can buy, but nothing is yet being mass-produced, and differences in efficiency are large. Try to see one of the machines in action before you buy one.

Before husking it is best to have your nuts at a proper state of dryness (though different husking processes are more or less sensitive to this

FIGURE 9-1. The cake box husker, shown assembled on the left, and with the top removed to show the hammermill elements on the right.

variable). In most cases the husking will be best if the husks are brown-dry, but not so dry that they crumble or crack before bending to let the nut out. If they *are* too dry, you can re-wet them using a mister, ideally letting the water be absorbed over the course of a couple of days. You don't want liquid water on the surface of the nuts/husks when you run them through the husker; that will result in wet husk fines (husk mud) collecting and sticking in corners of your husking and cleaning machines.

The easiest type of husking machine to build on your own is a hammermill-like machine, in which a rotating shaft spins rubber, plastic, and/or metal strips or chains through the husked nuts in a husking chamber. Our first iteration of this, which has found a somewhat broader implementation and which we still use for small-batch husking, is a set of chains and bungees attached to an electric drill, mounted in a plastic cake box, shown in figure 9-1. In use, the top of the cake box is filled halfway with unhusked nuts, then the bottom assembly with the drill is placed on top and latched on. The entire thing is flipped over and held at about 45 degrees from vertical and turned on. It is best to use a drill with a lockable trigger, plugged into a foot-pedal speed controller. Some batches will need a lot more power to get the husks off; others will need a lot less, to avoid cracking the nuts.

Hazelnut husker theory and design could well be a whole book in itself, but we'll keep it brief here. Here's a summary of some of the basic constraints faced in husker design:

- If you beat it too hard, nuts will crack. Usually these will be your biggest, thinnest-shelled, fanciest-looking, and most valuable ones.
- If you don't beat it hard enough, the nuts will stay in the husks.
- You don't actually want to get *all* the nuts out of all the husks, because weeviled, blank, and otherwise defective nuts often stick in the husks harder than the good ones do. A husker that leaves bad nuts in the husks is a good thing, because sorting afterward is much easier.
- Often a particular mechanism or machine tuning will separate husks from nuts in small clusters, but not in bigger ones.
- Husk moisture content has a significant effect on ease of husk removal in most cases. This can cause results to vary even when you're using the same husker on nuts from the same source.
- In some cases you can get better results either by machining harder/ faster, or by machining longer (particularly in small-batch huskers). Sometimes allowing a gentler husking process to go on for twice as long will result in a lot more of the big, expensive nuts making it through the husker uncracked.

 ## Cleaning Husked In-Shell Nuts

Once the nuts have been dislodged from the husks, the nuts and husks need to be separated. Technologies for separating and cleaning crop seeds are very highly advanced; but the knowledge of that industry has not yet been professionally tapped for the neohybrids. We will do that soon, when we have an extra ton or two to play with.

For the most part, the husker leaves you with:

- Husk fines.
- Husk leaves.
- Husked nuts, both full (filled with kernel) and empty.
- Unhusked nuts.
- Empty clusters: clusters of husks with the nuts removed.
- Larger trash (sticks).
- Cracked shells.
- Naked kernels.

Happily, husked nuts are more likely to roll and less likely to be blown by air than most of the other fractions. This means that a whole array of separation mechanisms can do something to separate whole nuts from the rest, though many of the current machines require either multiple stages or multiple passes for satisfactory cleaning. Blowers or aspirators in all their flavors, including with separation columns, can remove husk leaves, fines, most empty clusters, and even some cracked shells.

FIGURE 9-2. We constructed this shaker table using a children's playground slide. Its primary use is for cleaning nuts from debris after husking.

Shaker tables like the one shown in figure 9-2 can be good at separating cracked shells, empty clusters, unhusked nuts, and larger trash. This particular model is one we constructed using a small playground slide; it has an adjustable incline, and is hung from the support such that it can be shaken side-to-side; the table slots are much more sophisticated than you'd guess. When using such a table for separating husked nuts from the husks, set the incline fairly high, place a gallon's worth or so of the husker output near the upper end, and shake the table (we use both mechanical and hand shaking) while pushing the nuts and husks up onto the steep portion of the slide, then release. This motion allows the round nuts to roll down the slide to a nut-collecting bin while the husks stay near the top. After about a minute of repeating this motion, the top of the slide should have only husks, which you can push all the way up over the top backwards, to be deposited in a husk-collecting bin. Near the top you can see a shaker constructed from a repurposed hedge trimmer, which is crude but effective. Also note the pattern of slots cut in the upper portion of the slide. These allow fines to fall through and also provide a rough surface that you can use to help scrub husks off by hand, slow the nuts' fall, and direct their flow.

Screens are useful in cleaning. Depending on what you want to use your small nuts for, it can be sensible to remove fines and smalls from the crop at the same time, which we do in some cases with the rotary screen in figure 9-3.

FIGURE 9-3. The small rotary sizer made and used at Badgersett. Nuts go in at the right, then smalls come out the middle and larges go to the left end when you turn the crank. Hearing protection recommended. Putting more than 3,000 pounds a year through it will probably motivate us to build the next generation.

Cleaning and separating kernels from shell fragments after cracking is fairly similar to separating the husks, except that you need to use food-grade equipment. Most crackers give some percentage of kernels that are broken into halves or smaller pieces. Broken kernels are difficult to separate from the shell fragments using a plain blower or screening, so obtaining whole kernels is helpful for these types of cleaners. Other more sophisticated separation technologies exist, and should easily be able to separate broken kernels from shells.

The world industry standard for final-stage cleaning and quality control is the human hand and eye. Most commercial hazel operations in the world have a final human inspection process. Expect to do it this way, even if you've used some pretty fancy equipment in the earlier stages of cleaning.

Sizing

Why size? Tiny nuts are hard to deal with for human in-shell consumption and associated markets, though they're fine for oil and cooking. Eventually, cracking and cleaning technology for very small nuts will certainly improve, as it has for sunflower seed in the last thirty years. At the moment, however, hazel cracker technologies benefit from sizing, since they can produce a higher percentage of whole kernels if the nuts are separated into uniform size classes before cracking. The most common way is using a screen sizer.

The plain old rotary screen sizer, a small example of which is seen in figure 9-3, works well, but has critical details. This particular example constructed at Badgersett about 15 years ago has a door for loading in the nuts, as well as a hand crank, on the right side. Smaller nuts (and fines) fall through the screen, while larger nuts make it all the way to the left side. You need to have different screens, set for the sizing you want; changing your sizing parameters requires changing screens. It is possible to build a rotary sizer that can be fed continuously, not in batches, and can separate multiple size classes at once. Our current machine works just well enough, however, that we haven't gotten around to devising an improved version yet.

 ## Quality Control

Hazelnuts have an established world market, but the neohybrid hazels are new and a little different. If you hand a potential customer a hazelnut to taste (or if they've bought some of your nuts out of curiosity), there's a very good chance it'll be the first time they've ever eaten a neohybrid hazelnut. This is a *critical* first impression, and blowing it is extremely expensive both for you personally and for the neohybrid hazelnut proto-industry in general. This also includes all those ecological benefits associated with the crop. Selling bad nuts kills jobs, markets, and future ecosystems!

Neohybrid hazels come in a wide range of flavors, and they can be either good, strange, or bad in unexpected ways. Stringent quality control is a must. Toward this end, you need to be able to judge the taste of nuts. For serious judging of any flavor, some education and some preparatory steps before sampling are required. Read up on professional tea, coffee, and chocolate tasters. It's fascinating and you need to know.

When actually judging, first eat a not-too-spicy but satisfying meal. Everything tastes better when you're hungry; you'll have better judgment when you're sated. Too much spice in your meal will confuse things. Cleanse your palate in between samples with water, a mild cracker, or maybe a nibble of candied ginger (and some water).

It will be a decade, at least, before we at Badgersett will produce a hazelnut that we would dare to offer to any European company, except as a purely experimental experience. The reasons for that are quite complex. To begin with, the process of breeding a hazelnut that can be grown in the central United States has *not* included flavor as a selection criterion until quite recently. It couldn't.

Flavor is probably the most genetically complex trait breeders must deal with—there are certainly a minimum of 10 genes involved in adequate flavor, and another 20 to 30 genes or so responsible for exceptional taste characteristics. Add the fluctuations in flavor due to annual growing differences—drought, wet, cold or hot seasons, fertilizer levels—and you'll begin to appreciate how complex flavor can be.

When we began breeding neohybrid hazels, we didn't even have a plant that could reliably survive in the Midwest. The chances of starting at random and finding a climatically functional plant that also had a correct combination of the 10 to 20 genes responsible for noteworthy flavor are worse than vanishingly small; the world doesn't have enough space for the trial plantings (see chapter 12 for more about that). We had to first create a large gene pool of many plants that are reliable in our climate, then find those that produce reliable crops, and only later—once those characteristics were well established—search for flavor.

When we reached our fourth cycle of selection in the early 2000s, taste *became* a primary selection factor for the first time. We'd always measured it, but it could not be a primary factor until then. Bushes from those plantings are now coming into production, and taste will be a factor in the breeding program from now on.

One thing we know about the majority of our neohybrids is that their full flavor has not yet developed at the time of harvest. We believe it takes three to four months post-harvest for this, as long as the storage is appropriate; dry and not too hot (over 90°F, or 32°C). Cooler is better (we think) as long as they're still dry. At that point, many of them do indeed offer genuinely excellent flavor. Multiple attempts to perform controlled experimentation on how to reliably produce the best flavor have so far yielded only unclear results; we keep trying.

Some neohybrids exhibit excellent flavor. Some, definitely, do not. Which is not at all unlike the breeding pool for pure European hazels. In fact, we've documented unprecedented flavor types in some of our hazelnuts. Hazels that taste like walnuts or hickories. The most extreme is a series of types we can only describe as "aromatic-floral." They're going to be extraordinarily interesting down the road; our spectrum of available flavors is much broader than exists in gene pools containing only one species.

It's important to note that even though our hazel flavor is not consistent enough that we can confidently offer it to a European gelato maker, that doesn't mean it isn't good enough for high cuisine. Flavor diversity can in fact be a real selling point to some chefs and markets. The nut does still need to be good, though, and that is still on the uncertain side.

Hazelnut flavor is strongly influenced both by its current moisture content and by its moisture history—in other words, how long it has been how wet. To experience full flavor of the raw nuts, we judge moisture by how the kernel crunches between the teeth. It should definitely and distinctly go "crunch"; there should be little or no hint of "squish." At the same time, it should not be so dry that the kernel feels brittle when bitten. Once you've started chewing, hopefully you start to experience one of the following:

EUROPEAN HAZEL flavor is the characteristic "hazel" you taste in good-quality hazel oils and chocolates with hazel. The primary taste chemical giving rise to this distinctive flavor is filbertone, and in a high-quality nut it is underlaid by at least a slightly sweet and somewhat light, mellow flavor and feel. In most hybrid hazels the europeanoid flavor will not develop under normal storage conditions until several months after harvest, so far, and many plants will never show what we refer to as "good intense European" flavor.

AMERICAN HAZEL flavor is a little darker than European. It is distinctive and *not* the same as the filbertone-based standard European. It is often, but not always, associated with a more full-bodied feel, and sometimes a little bit of woody tang.

The most common flavor fault influenced by storage conditions is bitterness. A hint of bitter can add to the complexity of the bouquet, but in some cases a bitter flavor will develop as you chew, becoming an unpleasant aftertaste. The strength of this ranges from "I'd rather have eaten something else" to "spit it out *now*!" This is the taste defect that you most want to keep out of your product, and it's what most quality control guidelines are aimed at. As with any oil nut, plain old rancidity can be a problem, though hazels become rancid more slowly than pecans or walnuts. Most of the methods of grading work for both bitterness and rancidity. In general we expect moderately well-stored hybrid hazels to have potential for best flavor in the range of five to nine months after harvest; before this they probably haven't developed full flavor, and after this the taste is likely to start going downhill because the nuts have undergone too many temperature swings. The nuts can still be fine for most cooking or processing uses for an additional year or even longer, and refrigeration or freezing will greatly extend the window. We currently label graded in-shell nuts for retail sale with a "best by" date of six months after grading.

Woody tasting pellicle—the usually brown skin on the kernel—can be a problem with some individual bushes. A thick and fibrous pellicle leaves a distinct impression (and aftertaste) of chewing on sawdust. Genetics is the answer here.

Here are the descriptions of the shell markings potentially indicating kernel problems.

1. The black spot is a weevil oviposition scar. This may signify a deformed or a weevily kernel. Usually a deformed kernel tastes fine, though.

FIGURE 9-4. Examples of markings on the shell that may indicate a kernel defect. Numbered from 1 through 5 on the top row and 6 to 10 at bottom; details in the text.

2. This nut doesn't look too bad, but the non-hilum portion of the shell is on the dull side. Might be fine, might not.

3. This dark spot on the tip of the hilum is also a "sometimes good, sometimes bad" indicator. Whether these need to be removed will depend on the batch; crack five or so, and find out.

4. The completely dark hilum here is likely to indicate a bad kernel. Also, the slight fringe of white on the edge of the hilum can indicate a moldy kernel.

5. Dark hilum, might be bad.

6. Cracked nut—not necessarily a kernel defect, but should be removed because of possible contamination now, and likely insect infestation later.

7. Very dark nut; almost certainly bad.

8. Husk stuck on the nut; almost certainly empty.

9. Dark and uneven hilum coloring. This nut might have had its husk stuck and now be empty, or it may have a spoiled kernel.

10. Weevil exit hole. There is probably some kernel left inside, but you don't want to eat it. Livestock might; this is feed or fuel grade.

Kernels cracked out of their shells are usually easier to grade than in-shell nuts; you're looking right at the part that matters. Some of these defects have to do with the pellicle, or brown skin covering the kernel; others have to do with coloration of the kernel meat itself. Following are the defect markers for the kernels in figure 9-5.

1, 8, and 9. Scaly pellicle. These nuts are likely to have a woody aftertaste, although the nutmeat is probably fine.

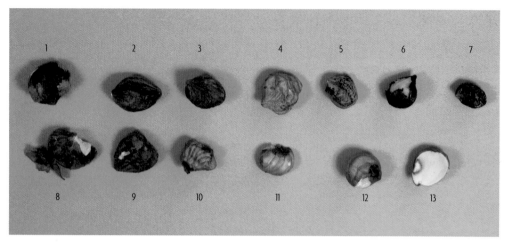

FIGURE 9-5. These kernels have one or more defects, though some are still edible. Numbered from 1 through 7 on the top row and 8 to 13 on the bottom row; details in the text.

2, 5, 7, and 9. Shriveled kernel. By itself this doesn't mean a bad nut, but it does indicate immaturity at harvest. These nuts are likely to go rancid more quickly, and their flavor is unlikely to develop fully.

2, 3, and 7. Dark pellicle; these nuts are likely to be rancid.

4 and 10. Misshapen nuts. Nothing wrong with this in and of itself, but it can be another expression of shriveling.

6. Flaking dark pellicle and yellowed meat are reliable indicators of spoilage.

10, 11, 12, and 13. The dark spots or pits on the pellicle may be associated with the white bad spot in nutmeat 13. This defect might not make you want to spit the nut out, but it almost always leaves a bitter aftertaste. Not food grade.

How to Grade Nuts

The existing hazel industries all have rigid, official, and legally required grading scales. However, they are almost entirely inappropriate for the neohybrids. Below are the processes we follow, and formal rules will certainly develop as production rises.

Removing defective nuts and grading by hand for human consumption is a process that currently works best with batches of 3 to 10 pounds. The diversity of the neohybrid nuts makes this more difficult than grading clonal nuts, so far. Work on a table or in a shallow round-bottomed bowl. You need to re-sample to make sure you got the defectives out, whether you're doing it in-shell or as kernels. To start with, let's define how to take a "sample."

1. Pick the worst-looking 10 nuts you can find near the top surface of the batch. Usually this ends up being the worst-looking nuts from a 200- to 300-nut sample of the batch.

2. Crack them.
3. Evaluate or taste them. For a sample to pass, all 10 nuts must be edible (you don't want to spit them out) and 9 need to be good (you're glad you ate them).

Here are the steps in the process of removing defective nuts and grading a batch:

1. First, pick a random 10-nut sample and evaluate or taste. This will give you an idea of whether the batch is worth grading or not. If there aren't at least five good nuts in that sample, there's a good chance you should put the whole batch in the fuel or feed nut bin at this point, rather than wasting time trying to separate out the good ones.
2. Remove nuts with obvious defects (weevil holes, moldy nuts, and the like).
3. Sample.
4. Remove the most defective-looking nuts from the entire batch. These are the nuts that have the darkest hilums, weevil holes, the blackest shells, or the most white on the hilum. Turn the batch over as you work, slowly stirring it so you'll eventually see all the sides of all the nuts.
5. Repeat steps 2 and 3 until the sample passes.
6. Take another sample of 10 nuts randomly chosen—not just the bad-looking ones, to make sure you haven't been missing some other defect. If this sample passes, the batch passes and is "good enough." If it doesn't, try to figure out if there is an indicator you were missing for the unacceptable nuts. If you don't find one, reject the batch.

When you complete step 6, the batch is good enough to eat—but of course this is just removal of defects, not an evaluation of whether the overall flavor and fill are of high enough quality. You want to avoid selling nuts directly to consumers if the flavor isn't sufficiently developed; you want the reaction to be "Hey, that's good," not "Hey, it's slightly better than cardboard." Similarly, in some cases the overall level of kernel fill can be low if the nuts were picked too early, or the season was extraordinarily stressful in some way. This can make it unsatisfying to crack them, in which case you probably want to avoid selling them in-shell to somebody who will be cracking them by hand.

Quality control can be aggravating. Sometimes a batch will need to be rejected due to an overabundance of one defect or another, or there will be a defect that you can't satisfactorily sort out. Sometimes it will be possible to lower a batch to grade B, meaning "suitable for use in cooking or further processing, but not for eating just like this." Sometimes a batch has so many quality problems that it's usable only as animal feed or fuel. Poor nut quality is much more likely in plantings with poor fertility, and will improve with better fertilization.

Since nearly all hazelnut defects are at least somewhat genetically determined, and there is a substantial amount of variability in the currently available genetics, it can be useful to keep batches of nuts separate based on either rows in the field or female parentage. This does require more bookkeeping, but it's very satisfying when you only need to reject one batch, rather than sorting through your entire crop—and you can get inferior genetics removed.

If you're having a bad day, or a bad season, for quality control, it can be cheering to go to the grocery store and buy a bag of commercially processed nuts—any type. Even if they're from a fancy source, there's a very good chance that at least once in a pound you will find a nut that you'd rather not have eaten, and occasionally one that you need to spit out. It seems to be the standard at least in the United States, and it is accepted for established nut types; perfection here is not attainable.

Cracking

Cracking hazelnuts by hand for personal use can be tedious, but is not the worst. At the very least you're going to need a hand cracker for field-checking nut quality and grading nuts pre-use. The best we've found are the antique HMQ nutcrackers, made by the company of Henry Quackenbush, the man who invented them (he also invented the extension ladder and a slew of other things).[1] You can find them on eBay without much trouble; our usual source is the tableware section of thrift stores. New ones similar

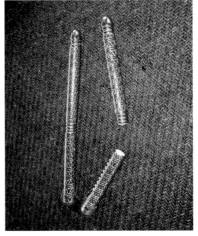

FIGURE 9-6. Genuine HMQ nutcracker, and one of today's knockoffs that was broken just cracking a plain ol' hazelnut.

to this seem to be always inferior—either the teeth aren't sharp, or the metal isn't strong enough, or usually both. Other fancy-looking new nut or crab crackers have so far been disappointing when we tested them for hazelnuts.

When you're cracking by hand, it's useful to know the tricks for easily getting the kernel to come out whole. There are many places you can grab the nut with the cracker where it will tend to break or smash the kernel when the shell cracks. Grabbing hold of the nut along the midline, slightly twisted as shown in figure 9-7, gives a whole kernel nearly every time.

Our experience with the machine crackers currently available is that they require you to size before cracking if you want a high percentage of whole kernels. We know quite a bit about cracking and cleaning hazelnut kernels but the depth of our experience with these tasks is substantially lower than it is with processing in-shell nuts.

If you are cracking for sale, you almost certainly need to do the task in a commercial-kitchen-grade facility, and the exteriors of the shells need to be sanitized first (because what's on the outside will touch what's on the inside when it's been cracked). Your cracker needs to be food-grade materials

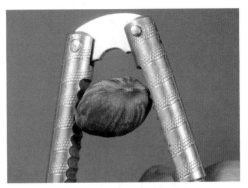

FIGURE 9-7. To obtain a whole kernel while cracking by hand, place the nut in the cracker so that the contact points are on the midline, as shown in the middle image, but twisted slightly so that one is closer to the hilum than the other, as shown at the top. This gives you the whole kernel shown at bottom.

and food-grade clean. Raw nuts are the largest source of food poisoning hospitalizations and deaths in the United States—take this seriously! It is true that the contamination risks for hybrid hazels harvested right off the bush are substantially lower than for nuts allowed to fall on the ground and be swept up, mixed with soil. However, there are still risks involved.

At this point you could not only be responsible for somebody getting seriously sick. In addition, any serious food safety issues with hybrid hazels will have highly detrimental effects on consumer opinion both of your crop and of hazelnuts generally. Be careful!

Sanitation via a bleach rinse (½ cup per gallon) is sufficient, and if you're quick about it the nuts will be dried back down quickly. Depending on your next step, the re-drying may not be necessary. If we're cracking nuts for our own kitchen, we almost always skip the sanitation step, because contamination from contact with soil is a vastly lower concern than with other nut crops. In cases where we've had to leave them on the ground in the field, or where rodents have made their way in, we do sanitize.

Roasting

All the talk about the taste of raw nuts notwithstanding, there are hazel researchers and growers who refuse to eat a raw nut: The taste of roasted hazels can be that much more intense. It's a great way to eat them; often nuts that are bland raw will be highly flavored once cooked. If you have access to a commercial kitchen, roasting hazels can be a way to add a lot of value to your product.

Before roasting, discard any kernels that are discolored or questionable in any way. Then spread the nuts in a single layer and briefly roast them at low oven temperatures (about five to seven minutes at 250°F, or 121°C), to give them the flavor we refer to as roasted. It is distinctively different from raw—more crunchy, more highly flavored, and with a more distinct hazel taste. Cooking just a little longer or hotter (five to seven minutes at 350°F, or 177°C) yields a toasted flavor. It is quite good; they'll taste like the best toasted almonds you ever had, but much of the aromatic hazel character will have cooked away. Since the process is so sensitive to temperature and timing, check a few kernels for taste once a minute before you do large batches. Cooking times and temperatures required are also sensitive to the moisture content in the kernels, so it's a good idea to monitor your first batch of the day carefully.

For many uses, you will want to put raw nuts into whatever item you are cooking; they will cook just fine. If you're doing in-situ roasting in a wet matrix like bread, you may need to coat the kernels with oil before mixing them in so that they do not become soggy while they sit in the batter before it dries out about halfway through the baking process.

Now you're ready to start looking up hazelnut recipes, to modify them to your taste, and to adapt them to the new flavor of these hybrids.

CHAPTER 10

Marketing

Ah, marketing. People tend to cringe at the word. Some loathe the idea of people manipulating them into buying something they didn't want. Some think marketing is incomprehensible (or are taught that), that they don't and never will understand it, and thus they don't like it. And this book is supposed to be about growing hazelnuts; what is a chapter on marketing doing here?

This book *is* about growing hazelnuts, but it's also about introducing people to a new crop. Many of you may plan to grow hazels only as food for yourself or to supply biodiesel for your farm. But some of you are thinking about growing hazelnuts as a crop—to sell.

Here's the thing: You will *never* make money growing hazelnuts.

But you *can* make money *selling* them.

There are farmers who love growing things and do so beautifully, but they're not enamored of the selling part, and would rather be anywhere else than sitting in a booth at a farmers market. It's not uncommon, and sometimes sad; too often those folks quit growing rather than face the burden of selling. If you're going to grow hazels as a crop, *someone* in your operation is going to have to sell them. Keep in mind that some people actually enjoy marketing. Find them. Start planning for it now.

This chapter is the barest introduction to the subject of marketing hazelnuts—enough to help you start thinking and planning. If we can get you that far, you will be miles ahead of most folks who planted an acre of apples, or two acres of walnuts, because they loved trees and the idea of growing something they might sell.

 ## Selling Neohybrid Hazels

Discussion of new hazel plantings always seems to focus on the nuts themselves—the nutmeats, in fact—and growing Food *is* one of the big reasons for doing this. It's the subject of this chapter. But if you're writing a business plan, keep in mind that neohybrid hazels also produce non-food products that can and ought to be an important part of your potential income. With the genetics currently available, more than 60 percent of the weight of the nuts you will harvest is shell. The shell protects the embryo against insects, birds, and microbes. Changing the kernel-to-shell ratio genetically is easy, but changing the ratio so that the shell still does its job adequately may not be, and requires considerable testing. European industry nuts have less shell and more kernel; we have breeding lines with kernel-to-shell ratios better than European standards, but they are not ready to release at the time of this writing.

You are going to have to do something with the nutshell, regardless of how much you want to focus on food. All nut industries—almonds, pecans, walnuts—have gone through this discovery, with the exception of pistachios, which have only been sold shell and all until recently. These nut industries began with a goal of reaching profitability by selling nuts as a luxury food. As they became successful and scaled up production, the piles of discarded shell grew until the nut processors were forced to do something about them. Only at that point, they discovered the profitable markets that had always been available for their shell. These days, it's even easier to find markets for nutshell; everyone familiar with any nut industry is aware of established uses, usually with an eye to talking that cosmetic company out of using crushed almond shell for facial exfoliant and into using, oh, say . . . much more exotic hazel shell. Badgersett has been fielding inquiries for 20 years from companies wanting to buy our hazelnut shell. Container-loads of it, and not for fuel uses. Yes, hazel shell

FIGURE 10-1. Hazel coppice wood. The value of your coppice wood will increase after the first coppice; it will come back straighter and bigger. Hazel rods from coppice are still used in Europe for many of the same purposes as willow: rustic furniture and the woven garden fence panels known as wattle, or hurdles.

makes excellent fuel and could be sold to pellet fuel makers, for example. But you will find there are much more remunerative uses, and buyers exist already (see chapter 11 for more about uses of shell).

Another product to plan to find a market for is wood. Neohybrid hazels have been bred from the outset with the expectation that they would be periodically coppiced—the entire bush cut to the ground, and the wood harvested as a product. We've started selling coppice wood products already, with enough success that our problem is not "How and where can we sell this?" but "How do we reliably provide enough to meet the demand?" If you don't make plans for selling your nutshell and coppice wood, someday you'll be scrambling to keep it from taking over your farm. From chips for fuel to artisan craft wood, it can make you money.

The Nuts!

The day will come when your local farmers market is no longer big enough for your hazel crop. The earliest larger-scale business or commodity markets for your neohybrid hazelnut kernels are unlikely to be in direct competition with established commercial producers; rather, we expect to be generating new markets. Remember, neohybrid hazelnuts have good flavor, but it's not identical to that of standard European hazels, and the flavor is not present in the kernels immediately after harvest. (See *Quality Control* in chapter 9 for details.) It is very easy to sell a modest amount of raw in-shell nuts at farmers markets; "local" is likely to sell very well for a long time yet. Dozens of neohybrid growers are already selling nuts this way. A big advantage for beginners of this type of direct marketing is that the processing costs are very low. Also, because nuts in the shell are considered a raw agricultural commodity, the liability exposure for the seller is quite low. Some states will allow you to sell the nuts pre-cracked, which adds value; some will allow you to crack nuts for customers only after the money has changed hands. With some work, you can build a direct sales business into a boutique-type operation if you can establish a reputation for top quality, and deliver it. Returns and satisfaction can be good.

Local specialties aimed at gourmet, gift, holiday, and other high-end markets are also within reach, waiting to happen. I have a collection of photos of hundreds of products already in existence that are waiting to be adapted and brought into production. Hazelnuts in honey is an easy one, and fancy; a dozen kinds are possible. Toasted hazelnuts ground with locally roasted whole-bean coffee. Hazel butter mixed with chocolate. Hazel brittle. Hazelnuts in 20 kinds of cookies. Cakes. Breads. Artisan cold-pressed hazel oil. Most of these already exist as commercial products,

You Have an "Unfair Advantage"

In the business start-up and planning world, advisers will ask you, "Okay, you've got product you want to sell—but what's your unfair advantage?" They're asking what your angle is; what makes your product stand out from the existing businesses you'll have to compete with.

When I teach the marketing section of our short course, I show a series of images of packages and their labels, and their prices. The series includes soybeans, corn, sunflowers, wheat, cheese, popcorn, bananas, and several other products. What they have in common is—they all call their products "*nutty!*" Some of them even specify *hazelnutty*.

In the food world, *nutty* is about the absolute best adjective you can use to sell your product.

Nuts are so attractive that the word is even used to sell, of all things, banana chips. Selling nuts amounts to all the "unfair advantage" you could ask for, all by itself.

in a small and usually regional way. Badgersett has experimented with making most of these using our nuts for years, and we've sold some experimentally at our farmers market (we sell out).

We also have a few products we want to try that we don't talk about yet. Keep your eyes and ears open and think innovatively. Chances are, you'll come up some product ideas we haven't thought of yet.

The requirements for value-added nut products are high, however. Food safety becomes a concern as soon as you crack the nuts; regulators will appear on your doorstep. You need to learn what you can expect. States regulations vary; federal laws in theory don't, but in practice may. Information about regulations is much easier to find these days than it used to be. If you go to an established farmers market, they will almost certainly have a person who is there to teach you—they have to be in compliance, or they'll be shut down. Many states will have a regulations ombudsman in their agriculture department whose job is to walk you through the process.

For human consumption, nutshell has to be thoroughly removed—very close to 100 percent. Most nuts can support dangerous microorganisms if mishandled, anywhere from harvest to the retail shelf. There are allergy concerns; are your labels clear and legal? Hazelnuts belong to the group of nuts that can lead to serious reactions or even death among people with severe nut allergies. (The same is true about almost every other food, of course.) In our experience, adults who have such allergies are meticulously careful about what they put in their mouths, and also keep very close watch on affected children. This is a concern, but one shared by nearly all food producers, and awareness at every step in the food system is very high these days, from grocers to wait staff, and allergies should not be a barrier to sales.

If you want to be successful, the flavor of your hazelnuts will need to be not only good, but consistent. On a larger scale, you will want to be able to assure customers of a steady supply, which means you need enough land in production to provide the supply for any new market, as well as your old ones. High-end retailers may drop you if they invest time and shelf space to develop customers for your product, and you then experience a break in your supply. Processing and production of high-end products, such as boutique chocolates, will be expensive.

If neohybrid hazels are to reach their full potential—becoming Food for cities and replacing naked soil annual crops with functioning ecosystems—we will need to develop much larger, commodity markets. My favorite historical model is the soybean industry.

In the early part of the 20th century soybeans were a new alternative crop, with no established markets. Farmers discovered they could grow them successfully. They started to work hard on developing new markets and new products. Today, chances are that wherever you go to shop or eat, you can easily find 20 different items that contain soy products. When neohybrid hazel growers reach the point of collectively producing a sufficient volume of nuts, those nuts can be drop-in replacements for soybeans. Any industrial product made from soybeans can be made from hazels, I promise you—and quite a lot more. Hazels actually make a better chemical feedstock than do soybeans; their oil and protein are better, and they're non-toxic (see *Resources* at the end of this book for hazel chemistry). In the near future, thanks to the crop dynamics of the neohybrids, there will be no tillage expenses, no annual planting with those risks, no sprays, and no costs for fossil fuels or machines to carry out those jobs. Neohybrid hazels will be able to produce oil and protein more cheaply than soybeans. That's the point at which soybean farmers may start to switch to hazels on a large scale. The soybean industry need

Price Per Pound?

It's the million-dollar question for any product—what price to charge? You can look up the current commodity price for hazelnuts on the dock in New York, or Hamburg, but that information isn't helpful if you're selling your nuts to a local chef. On a small scale, your best bet for setting prices would be to go to your local grocery store and check the prices there. Then add a bump because of the extra value of your product as locally produced and/or organic. The reality is that in the early years for this crop, you may be the one who is setting the price. An excellent hint: If your customers don't complain it's high, you truly need to set it higher. This may sound rude or greedy, but it's straight truth. If they don't complain but walk away without buying, lower it.

A Few Good Marketing Principles

LESSON 1: Don't waste your time fretting about what makes sense. It's entirely likely that what will sell your product will not make sense. But it will sell your product.

Marketing is learning to understand some quirks of human behavior that have been extensively studied by the people who place the shampoo and the potato chips in precisely chosen spots on particular shelves in Target and other big-box stores, with the label on the chips printed in red and the shampoo label in turquoise. The goal is to be sure the products you want to sell, because you've put your lifeblood into them, will sell. And sell fairly. And in quantity.

Yes, you need to learn about it—if you don't, you *will* be ripped off. Whole books have been written, some as humor, about great marketing blunders. Fools and charlatans thrive among marketers, too; you need to have enough information to recognize and avoid them.

LESSON 2: Listen to your instincts—they're worth listening to, and they might well be right. If you're trying to sell a new product into a new market, it does make sense to ask advice from experts. But . . .

LESSON 3: Beware people who tell you they are experts. The most knowledgeable people I know, in all endeavors, usually insist they are *not* experts. Meanwhile the ones who insist loudly that they *are*, are the most likely to be downright charlatans.

You can find self-deluded experts all too easily. A classic marketing blunder was made in the apple industry some decades ago. Professors (who were not trained or qualified in psychology, or testing, or marketing) did "marketing tests" that "proved" that the American public wanted only big, red, pointed apples, with maybe a Yellow Delicious occasionally. According to those experts, under *no* circumstances would any American consumer ever buy a green apple. So New Zealand planted, developed, and marketed the grass-green Granny Smith, making many millions of dollars shipping them from their side of the world to ours. This continued for years until, finally, American apple growers woke up, ignored the professors, ripped out acres of Red Delicious apples that weren't selling because people loved Granny Smith, and started planting Granny Smith. Those professors had not the slightest idea how to do valid tests, and didn't know that they didn't know.

LESSON 4: When someone tells you marketing is a mystery—run. It's a science, with an enormous amount of very intelligent and valid research behind it. This research is applied to determine the exact best placement for every package, the exact best slant and color for every font, the exact best weeks of the year a certain product should be displayed on a certain shelf. All of these factors have been studied, and tested, and are being monitored right this second to measure sales results. It's a science. Science is never infallible, but it beats the heck out of guessing and hoping.

not suffer in this process; farmers can just switch crops, and the local soybean-crushing plant can be modified, very slightly, to be a hazel-processing plant. Everybody will keep their jobs—and the air and water will be cleaner.

We have the genetics to move in this direction now; what we need are growers, in many regions, finding and developing regionally adapted neohybrid strains with a wide genetic base. Oh, and "product." I have been repeatedly amazed by how many people are eager to develop machines, retail products, and growers associations before there are more than a few pounds of local nuts to work with.

The processing requirements to get into non-food commodity uses of hazelnuts, such as biodiesel, are dramatically less than those needed to get into top human consumption markets. Price per ton would be less, but production cost per ton will also be less, making such markets potentially quite profitable.

The High-End Markets

When I was doing taste-testing with the chefs and foodies at a Minnesota food show in 2002, I of course made the rounds and talked to the chocolatiers. And this was high-end chocolate: It was almost Valentine's Day, and one of the truffles I bought for my wife was sprinkled with 24-karat gold leaf. I found it hard to keep these chocolatiers' enthusiasm connected with reality. They were all very eager to work with local hazels; they knew they could sell them. They tasted the nuts with no concerns that our hazels did not taste like the hazels they were accustomed to. "I think those would be fun to work with!" I heard more than once. That, and "Let me know when you've got nuts!"

About half the attendees at this show were chefs, and I carefully pointed out to them that each individual nut may have a different flavor. The chefs had the same response as the chocolatiers: Half of them quickly said they'd enjoy cooking with nuts with variety.

The high-end markets are open to neohybrids. What they will demand is clean, near-perfect nutmeats—they're not going to crack and sort their own—as well as a supply they can absolutely count on. I don't think you need enough to supply them 12 months a year; these are products where holidays are important, and if you can supply a top market for the duration of one holiday, they'll buy cheerfully. Products available only for a short time are their bread and butter. Still, if your entire harvest would last them only three days, they're not likely to be interested.

About 80 percent of the existing world hazel crop is consumed in Europe, most of it in some combination with chocolate. That's good information, but it may be of little use when it comes time to start selling your own crop. Those markets are old, established, and finicky. They have little tolerance for ingredients that vary from their standards and practices. The Oregon/Washington hazel industry has had little success attempting to export its hazelnuts to Europe. The Europeans' reaction to the nuts was: They don't taste the same; the shape is not the same; the pellicle (nut skin) doesn't behave the same; and they're too big to fit in the candies we make. Oregon/Washington are now exporting almost half

FIGURE 10-2. Philip passing out cracked and sorted raw neohybrid hazels at a food show, testing chef and consumer reactions in detail. (They liked them.) Photo by Brett Olson.

their production, but their main customers are China and Vietnam. These are new markets, without set expectations, and where nuts from America are seen as special.

The Ferrero Company from Italy has been brilliantly successful in marketing its candies in the past couple of decades. Their success is based on two things—intelligent marketing and an insistence on maintaining quality, in both the chocolate and the hazelnuts. It delights me no end that I can now walk into any supermarket or convenience store in the United States and be certain that I can find Ferrero Rocher candy for sale in the checkout lane—and if I am in the mood to buy a few I can count on the whole hazelnut inside being a really good hazelnut. I can't remember buying one and being disappointed in the nut, and that is not the case for some other brands.

Ferrero has been selling a lot of hazelnut chocolate and Nutella in North America. Shipping chocolate across the world during the hot months of the year can be risky and expensive. Moving manufacturing closer to sales makes good sense. In fact, Ferrero has expanded their manufacturing to North America; in 2006 they opened a large new production plant in Ontario, and recently expanded it. The great majority of the hazelnuts they use at this plant are still shipped from Turkey, however. Ferrero buys a few hazels from the Pacific Northwest, but in general the company has the same reaction that German chocolatiers did: Those nuts don't fit their business. They know how to sell hazelnut chocolate, and want what they've always used. Ferrero is trying to convince farmers in Ontario and New York to grow standard cloned European hazels. (I'm not holding my breath. Smart people have been trying to grow those genetics there for well over 100 years, with no commercial success.)

In the business world, this is a classic opportunity. There is a new commodity supply here—which old processors won't use, because it's slightly different? Hmm. Start up a business to use the new supply—in a new way. That business model has been working for millennia. Let us know when you have nuts to sell. Badgersett will buy them, if you aren't ready to start your own hazel-brittle company just yet.

CHAPTER 11

Co-Products and Their Value

A major strength of neohybrid hazels as a crop is the valuable co-products they produce in addition to the edible nut kernel. Sale of the co-products can rival nut sales as a source of revenue and seriously improve the bottom line. Although we call them "co-products," they are by no means a minor part of the system; there may well be years when non-food products are what make you money.

At the same time that the bushes are building nut kernels, they are also producing wood, leaves, husks (aka involucres), and shells. Each of these has value either on the farm or as a commodity in its own right. The value of some of these products has been recognized and realized for centuries. Others have known value, but developing a market for them will have to wait until production as a whole is increased substantially, because they are derivatives of fractions of the plants (for instance, isolated compounds from the husks). We've included descriptions of these products to stimulate your creative and entrepreneurial juices.

 ## Wood

Management by coppicing to rejuvenate plants and maintain their nut production at optimal levels includes harvesting all the aboveground wood on a regular basis, usually every 8 to 12 years. The optimal coppicing interval for your planting will depend on whether your goal is to maximize wood production or nut production. With some of our plants reaching 18 feet in height and spanning 10-foot aisles between rows, the amount of wood harvested per plant is substantial. We've noted wide ranges in vegetative growth characteristics from different plants: wood density, amount of growth made, strength characteristics, number of stems, and breadth of the plant crown. This indicates that you could seriously impact the wood quality and quantity produced by selecting for your desired traits, eliminating the less desirable plants, and propagating, by either seed or division, from those kept to fill in the holes. At Badgersett, the focus is on nut production, and we have not generated separate breeding lines for wood traits.

Hazelwood is usually strong, flexible, and tough, but not particularly rot-resistant. Items made with hazel are likely to last a long time, as long as they aren't left exposed to the elements and/or in contact with the ground.

Beginning in 2015, Badgersett will begin splitting our neohybrid gene pool into two separate groups—those that make tall plants and thicker stems like European hazel, and those that remain fairly short, with thinner wood, like the two North American species. These growth forms are proving reasonably heritable (approximately 60 percent) even with an unaltered neohybrid pollen cloud; with both parents of the same form, heritability is around 85 percent. You'll be able to choose which type you plant, which will make harvest easier. Your wood harvest will also be more uniform, which means easier to use and market.

Biomass for Energy

Like other members of the birch family (Betulaceae), and especially fellow members the subfamily Coryloideae[1] (*Carpinus*, musclewood or hornbeam, and *Ostrya*, ironwood), hazels have dense wood that generates high energy per volume and has relatively low ash content. The one wood density statistic uncovered for the genus was 45.9 pounds per cubic foot. Although the statistic is for Turkish hazel (*Corylus colurna*), the other species and hybrids are likely to be similar. This ranks hazels just below white oak (47.2 pounds per cubic foot) and above red oak (44.2) and white ash (43.4).

The major stems can be cut into lengths for domestic woodstove use, or chipped for larger scale or auger-driven biofuel furnaces. The chipping option is also useful for essentially converting the twigs into pellets with none of the mechanical pelletizing or condensing required by some other biomass fuel stocks such as grasses or corn stover.

Outdoor wood furnaces could easily use hazel "faggots"—a bundle of stems bound together and cut to length. The faggot was a standard fuel in most of Europe through most of recorded history.

Rustic Furniture and Crafts

The branches of hybrid hazels tend to grow long and straight with little taper from top to bottom, especially following coppicing. The flexibility of these branches varies among genotypes, but in many cases a green 8-foot length can be bent so the ends touch without breaking or splitting; dry wood is steamed before bending. This makes them prime candidates for use in many types of bentwood furniture or craft construction. A fair number of artisans are doing this in England, where these are ancient traditional skills and coppice wood is readily available, but with limited mature neohybrid plantings the field is wide open for North American growers.

FIGURE 11-1. A new hazel hurdle fence, about one week to make. Once used as movable sheep fence, the use of hazel hurdles in modern England still protects privacy and provides windbreaks. Hazel stems have also been the standard framing for thatched roofs and the wattle in wattle-and-daub wall construction. With the increased interest in eco-friendly building materials, this could be a business opportunity for growers. Photograph by Steve Brown, Brown's Hurdles, Dorset, UK.

In England, where hazel coppices have been maintained for centuries, artisans still employ the traditional techniques for the weaving of hurdles (short fences for sheep exclusion and privacy) and ornamental pieces. As with bentwood furniture, this craft can flourish only when adequate coppiced stems are available.

Thicker stems are one of the most preferred woods from which to make quarterstaffs, the 7- to 8-foot-long sticks used in traditional European stick fighting. Picture Robin Hood and Friar Tuck. With nearly 30,000 members of the Society of Creative Anachronism, there is likely a Renaissance festival in a city not far from you—and a market.

Smoke Wood

Hazel wood can be used green or dried to produce smoke for flavoring foods. Its flavor is not as intense as hickory and can be used to enhance milder flavors such as fish. The flavor imparted is similar to that of alder, its close relative, which is the dominant smoke wood used with fish in the Pacific Northwest. Hazel has been sold with a variety of Badgersett woods at farmers markets in two neighboring communities. We found that smoke wood, in general, has promise as a stocking stuffer gift for foodies and as a souvenir for tourists. Some customer education needs to be done before sales of hazelwood will rival those of hickory or apple, but a local chef in search of a novel dish or two could be enlisted to help with this.

 ## Shell

Shell makes up a little over half of the weight of an intact nut. The data from Badgersett's entire database ranges from 50 to 80 percent shell for a given plant, with the current median value for Cycle 4 data being 63 percent shell/37 percent kernel. Shells provide necessary protection to the kernel from insects, birds, and rodents. They also protect the kernel from the environment in terms of spoilage from microorganisms and fungi, and oxidative degradation (rancidity) of the component oils. Although few growers will likely consider this, it should be noted that for fuel use, a generation or two of selection for increased shell thickness and percent of whole nut weight would increase the shell content significantly. We've been selecting in the reverse direction—for greater kernel percentage.

Biomass

The shells are very dense woody tissue consisting of cellulose, hemicellulose, and lignin. A study done on Badgersett nuts found them to be highly condensed sources of energy.[2] The intact nuts, including the high-energy oil from the kernels, produced 8,800 BTUs per pound. The shells alone produced 7,900 BTUs per pound. The amount of ash produced was only 1.45 percent and 1.64 percent, respectively; exceptionally low for biomass fuels. The whole nuts and shell fragments both flow readily and can be easily used in automated-feed biomass furnaces; we also have reports of shells being used in unmodified pellet stoves. We have no reports, so far, of neohybrid producers with enough product to sell commercially for this purpose.

A couple of notes on use in domestic woodstoves. For complete combustion, it is important that aeration be maintained—for instance, by use of a screened rack like those for pellet stoves. Feeding too many nuts at once can result in smothering the fire with too much combustible gas for the available oxygen, as the oil will volatilize quickly. When burning whole nuts, it is also advisable to do so in moderation to avoid overheating of unmodified woodstoves. The coals from the shells are very long lasting. Whole nuts burned at lower temperatures can result in oils condensing inside chimneys, producing heavy creosote and an increased risk of chimney fires. We use non-food grade nuts in our woodstoves, carefully. Stoves designed to burn hazelnuts would be a great idea.

Charcoal/Biochar

Charcoal contains more energy per pound than intact wood or nutshell, since it is essentially a purified fuel. Considering how concentrated the fuel energy was to begin with, this makes for a very powerful fuel that is especially useful in the forging of carbonized steel for artisan sword and knife blade production, which require temperatures of 2,000 to 2,100°F (1,093 to 1,149°C).

Activated charcoal is charcoal that has been treated (physically or chemically) to increase its adsorptive capacity two- to threefold. Many toxic molecules will bind to the charcoal, making it an effective treatment. Medical-grade charcoal is used internally as an antidote to poisoning, including drug overdose, and topically in poultices and compresses. This is also the material used in masks and other filters for air and liquid purification. Production of activated charcoal from hazel shell or wood could produce a significant side income. The amount and quality required to be able to attract the interest of the charcoal industry might make this proposition more attractive to a regional growers' group.

Finding a definition of *biochar* is contentious, but mostly it differs from charcoal in the intended end use of the product as a soil additive. Biochar researchers are still arguing about the temperature and duration of pyrolization (use of heat to cause decomposition of organic matter) used in order to optimize the char microstructure for its intended use. When biochar is used as a soil amendment there is a whole litany of benefits to the production system. The biochar particles improve the soil structure and fertility by providing habitat for soil organisms, improving both water retention and soil aeration, reducing compaction and compactability, and improving fertilizer efficiency by providing a site for the plant nutrients to adhere to until they are extracted by the crop roots.[3] This also reduces leaching and groundwater contamination.

In addition to the productivity of the agricultural system, however, is the fact that burying this very persistent form of charcoal constitutes serious carbon sequestration. When produced correctly, biochar is nearly pure carbon, and is almost invulnerable to microbial attack. Substantial biochar use has been proposed as one of the very few measures capable of actually *reducing* greenhouse gases in the atmosphere and thereby mitigating global climate change.[4] And you can easily sell it, today; around $10 per pound at the farmers market or $500 to $1,000 per ton in bulk.

Abrasives: Industrial and Cosmetic

The shells of hazelnuts are extremely dense, and their hardness approaches that of quartz sand. Where hazels are grown in abundance there is an international market for the shells as industrial abrasives for polishing metals or cleaning buildings. Like walnut shells,[5] they can also be ground and sized for use in cosmetic exfoliants and scrubs.

Truffles?

You may have heard that people are planting hazels inoculated with black truffle, and others, in North America now. So far, we haven't been able to get truffles to grow compatibly with neohybrids, however. We've worked closely with two of the oldest truffle growers in the United States, and at this point, it seems likely that European truffles act as a pathogen with North American hazel genetics. In Europe, hazels that grow truffles produce few nuts—their requirements are different. It does not seem a promising direction.

Particleboard

There already exists a healthy international market based primarily in Asia for hazel shell to use in the production of particleboard, in order to make a tougher end product. As the US forest product industry struggles to compete, it could conceivably make good use of a local resource.

Landscape Uses

Hazel shell biodegrades slowly, which makes it useful as a mulch material for planting beds and paths. Shell fragments could serve as a driveway surface in place of gravel. The sharp edges on the fragments make hazel shells advantageous in slug control,[6] but also mean that such mulch should not be used in areas where bare feet are likely to tread. The shell components can be recycled through composting, but the shells may take years to decompose. The rate of decomposition depends to a large extent on moisture levels and amounts of nitrogen available, but is much slower than herbaceous landscape debris or other woods under the same conditions.

Shell fragments can also be very effective as a non-chemical and biodegradable means of increasing traction on ice and packed snow for foot or vehicular traffic. We are excited about giving this use a serious trial in our laboratory . . . rural Minnesota.

 ## Involucres/Husks

Because the nuts of neohybrid hazels typically do not fall free from the husks at maturity, the husks become part of the harvest. The husks can make up 50 percent or more of the harvest weight on a fresh basis, though the range of percentages is substantial. On a dry(er) weight basis, the husks can still contribute a large portion of total harvest; the nuts dry out, too! They have potential value as animal feed and a source of biologically active chemicals for the pharmaceutical and agricultural chemical industries.

Animal Feeds

Although no one we know has measured the nutritional composition of hazel husks, at least one grower reports that the palatability of the husks for hogs is high, whether fed green or slightly rehydrated after drying. Anecdotally, the author Laura Ingalls Wilder wrote of enjoying the slightly acid flavor of hazel husks. Despite extensive sampling, we have only rarely located that trait in any of our hazel husks—acid yes, pleasant no.

Prospecting for Novel Chemicals and Pharmaceuticals

The husks appear to be instrumental in reducing predation and egg deposition by various insects. Different plants can have very different husk chemistries. There are also differences in insect damage in plants where the involucres cover different percentages of the nut. The types where the husks cover just the base of the nuts can show much higher incidence than the "filbert" types, with the majority of the entire nut enveloped by the husks. The husks vary for uncertain reasons too; on some bushes the husks will be immaculate at harvest, with no bug bites anywhere, while the next bush may have husks badly bitten. We track these data.

The discovery of the active chemicals in the husks could result in advances in insect control either as a botanical product or as a basis for

FIGURE 11-2. Glandular hairs tipped with a drop of liquid that is full of active chemicals occur on leaves, husks, and current-year stems of neohybrid hazels.

industrial synthesis. Whether the compounds are a toxin or a repellent (or both), new products resulting in reduced crop damage are very valuable and can result in monetary gains for you, the producer, both from the proprietary discovery aspect and from the sale of husks as the substrate for biochemical extraction.

Note: Production of many such chemicals may depend on which pests are present, and have been present in the past, since they are often produced only when elicited. Whether the compounds are very specific in their action or protect against a whole class of pests (say, insects) will have to be determined by appropriate testing. The potential is also good that the husks, and even the shells, contain antibiotics, fungicides, and/or pharmaceuticals for human and animal treatments.

Sub-Food-Grade Nuts

A certain percentage of your crop will not be up to standards, yours or the industry's, for human consumption. The components of the nuts are such that this needn't be a loss. Two notable avenues for use of sub-food-grade nuts are as animal feed and the production of biodiesel.

Animal Feed and Nut-Fed Specialty Meats

Specialty meats command a premium price, and hazelnut-fed pork has proven a winning combination. The hogs are fed a combination of grains along with the nuts for the final one to three months of their lives. The flavor of the fat is altered—and whether it is the heritage breeds genetically inclined to have large fat deposits or the trimmer modern versions, the meat is in demand. A few years ago, the asking price at the Eastbank Farmers Market in Oregon for fresh pork fattened on hazelnuts ranged from $4.25 to $8 per pound.[7] This market can be sufficiently lucrative that one neohybrid grower we know has shifted their entire crop to feeding their specialty pork.

Other animals, such as poultry, can benefit from hazelnuts in their diet. In formulating animal diets, hazelnuts contribute very high-quality fats, 15 percent crude protein, and 9.7 percent crude fiber—which is more than is provided by soybeans. Hazelnuts that have been cracked but not had the kernels separated from the shell are a treat for chickens to pick through.

Biodiesel

Badgersett has been working in this direction for over a decade. For our 2007 field day, we used raw hazel oil extruded from whole nuts (shells included) pressed by University of Minnesota professors Paul Porter and Derek Crompton (and allowed to settle for a day) to power our John Deere diesel.[8]

FIGURE 11-3. Brandon is about to pour a jar of the raw hazelnut oil into our tractor fuel tank at the 2007 field day.

Studies of the physical and chemical properties of hybrid hazel oils found that they are superior to soybean for use in biodiesel production by a couple of measures: The oxidative onset temperature (an indicator of thermal stability) is 97 to 108°F (36 to 42°C) higher than soybean, the cloud point is lower (−22.3 to −26.8°F, or −12.4 to −14.9°C, versus −17.8°F/−9.9°C), and the kinematic viscosity was also superior.[9] This is in addition to the greater production of oil per acre—of almost two-fold—than soy, even using yield data from the field of older Badgersett Cycle 2 neohybrid genotypes planted at Arbor Day Farm.[10] Using hazel for biodiesel has yet to be done on a commercial basis.

The uses of hazel co-products will continue to grow and evolve as more acreage is planted and becomes productive. It is not just a single food item that is being developed but a whole cluster of related products and industries that can bolster the economic health of producers and rural communities.

CHAPTER 12

Neohybrid Hazels—
Beyond Mendel

By now, you've learned enough about neohybrids to understand that they are genuinely different from all other kinds of hybrids. What is going on inside them, at the cellular and DNA level, is distinct indeed. And thus, we need you to understand that while you *can* plant seed from your own neohybrid hazels to expand your planting, unless you do so knowledgeably the outcome will likely be very poor. There will be no return on your investment and time. Years you cannot recover will be lost.

Unlike hybrid corn, which has minimized genetic diversity, neohybrid seedlings are specifically highly diverse; in fact, neohybrids are greatly more diverse genetically than wild species are. Simultaneously, they are also capable of immense heterozygosity; again, more than any wild species could be. These statements are not guesses or opinions; they are mathematically provable. Yes, we've done the math. The genetic variation available to us in this hybrid gene pool is truly enormous; the genome is certainly larger than the entire world corn genome. In past publications, I've stated, "Perhaps 10 times larger; perhaps more than that," knowing it was much more, but that no one would believe me. Brandon is more of a math enthusiast than I am, so I asked him to run the numbers. According to Brandon's calculations, even with very conservative genetic diversity estimates for hazels, it looks like this hybrid swarm has 10^{150} or so times the diversity of the world corn genome. The exponent *150* is not a typo. When you mix three species multiple times, the number of possible combinations—many of which have never occurred before—explodes.

Your mind may be boggled by the thought of such huge numbers, and by terms like *heterozygosity*, and that's okay. In this chapter, we are going to help you understand enough principles of genetics so you can become part of the larger effort to continue selecting and improving hazels.

 ## Today, and in the Future, *Farmer = Geneticist*

How will it help you as a grower to delve into this darn genetics and hybrid-this-and-that stuff? First, understanding the genetics is important in deciding what planting stock to buy (and avoiding the wrong planting stock) and also in making crop management decisions. If that worries you because you don't think you can master the genetics, it shouldn't. Nearly all farmers today—whether they're raising beef, dairy corn, or beans—*are* geneticists. They have to be. I know farmers in my area who are better practical geneticists than some professors of genetics I'm acquainted with. Go online and look up "dairy bull tests" or "new corn hybrids." You'll find pages full of acronyms and tables that will be entirely incomprehensible to you, but most farmers can read and interpret them at a glance.

You *can* learn to interpret data about hazel genetics and inheritance of traits. You don't have to know how the genes work exactly, but you do need to know how well traits can be passed on. Once you master the concepts, you'll have the tools you need to move your hazel planting forward. It may help to keep in mind that Darwin worked out a tremendous amount of how inheritance works before anyone knew genes existed, let alone DNA.

The other reason to understand and work with hazel breeding is that the world needs Food, with the capital *F*. Food to feed cities; crops that will grow and feed hungry people regardless of droughts, floods, downpours, and locust plagues. And crops that do not ruin the land and water.

We are working to make the neohybrids part of the complex web of tightly interrelated species and organisms that are humanity's symbionts. Species such as rice, maize, cassava, and chickens. Our symbionts number in the hundreds of species (including ones you'd rather not think about, such as black rats, bedbugs, eyelash mites, syphilis spirochetes, and polio viruses).

Without symbionts, civilizations would quickly die, as would most individual humans. Many of our symbionts would likewise vanish without a dense population of humans to support them. As a tiny example, most breeds of sheep and dogs would not survive without humans to care for them. A few would subside back into wild species, but Chihuahuas and Jersey cows are not going to make it.

It's a valid philosophical question whether we created our symbionts or they created us, but the point is that we live with, and by, fellow creatures that we have changed and that have changed us. Did we "create" domestic sheep, or did they create our cities? Wool was critical for millennia in keeping humans alive through the seasons, and helped us survive in parts of the world otherwise not open to us. Meat was important, but possibly more important was cheese: highly concentrated nutrition derived from grass, which in its natural state is useless to us for food. And cheese could be stored over winters, when food was difficult for humans to find. It was a long process, moving in small steps over centuries, and highly entangled with other species as well; dogs are critical for most sheep–human cultures, for instance, to keep wolves away and keep flocks together. When you start trying to tease apart the relationships of our symbionts, you find you cannot. We depend on one another; it's a matter of life and death, for all of us.

Some of our old symbiont allies in the world have reached a point of development where they are becoming toxic to the earth—potentially lethal to humans and our companions. If we continue to plant more grasses for seed, to rely primarily on them for our food, and to feed them to livestock in concentrated animal feeding operations (CAFOs), the grasses may just kill us all.

It seems likely that if we are to prevent our cities and civilizations from collapsing, we must have some new symbiont crops—something truly new, truly different, from the ones that put us in this situation. Neohybrid hazels have that potential, but there's more breeding work to be done. We need a more diverse population of neohybrids, with different characteristics that will allow them to thrive in a very wide variety of regional climates and conditions. As a comparison, think about how many different kinds of apple there are. Easily 500 or even 1,000. How many kinds of potatoes are there in the Andes? Again, hundreds, at least. All with useful attributes, and all created by farmers who paid attention to their crops.

 ## Hybrid Swarms, a Natural Phenomenon

As we use the word, a *neohybrid* is a cross containing more than two species, with new crosses carried out and selected for multiple generations—at least six. We're not aware of any other breeding efforts that have carried out the crossing process to this extent. Hybrid oil palms are usually crosses between two species, carried out only one or two generations beyond the initial cross, most frequently with subsequent crosses being back to one of the parent species, which reduces genetic diversity. Hybrid hazels previously in existence are the same: crosses between two species, usually followed by crosses back to *C. avellana*; a few are now being made that go one generation beyond.

I first heard the term *hybrid swarm* when I was in graduate school, taking a course in ecological genetics with Dr. David Merrell. The class correlated population genetics, ecology, and evolution. While we studied species hybrids and populations, the term was not in use yet. When I first started talking about neohybrids, about 15 years ago, I found that even academic audiences were rarely familiar with the necessary terminology, or with the actual genetics involved. More than once I used the phrase *hybrid swarm* while addressing a large group of PhD biologists and got a lot of blank looks.

Today the term is coming into wider use, as is its derivative *hybrid speciation*. Recently even Wikipedia has added definitions of these terms, and I find those definitions unusually lucid and helpful:

A hybrid swarm is a population of hybrids that has survived beyond the initial hybrid generation, with interbreeding between hybrid individuals and backcrossing with its parent types. Such populations are highly variable, with the genetic and phenotypic characteristics of individuals ranging widely between the two parent types. Hybrid swarms thus blur the boundary between the parent taxa.

Hybrid swarms occur when the hybrid is viable and at least as vigorous as its parent types, and there are no barriers to crossbreeding between the hybrid and parent types. Swarms cannot occur if one of these conditions is not met: if the hybrid type has low viability, the hybrid population cannot maintain itself except by further hybridization of the parent types, resulting in a hybrid population of low variability. On the other hand, if hybrids are vigorous but cannot backcross with parent populations, the result is hybrid speciation, which, aside from the contribution of new hybrids, evolves independently of its parent types.[1]

While the majority of known naturally occurring hybrid swarms are made up of two parent species, there is no barrier to adding more species to an existing swarm. Doing so increases potential genetic variation hugely. That variation is the basic material for the breeder to work with.

The potential for hybrid swarms to result in new species became clear to me while I studied with David Merrell. At the time, this path to speciation was controversial. Other evolutionists wanted to push their own ideas for how new species come to be. Now, following the development of DNA analysis, it has become widely accepted that hybrid swarms can result in new species in all taxa examined, but especially in plants. It will be "discovered," in another decade or two, that hybrid swarms are one of the most common ways that new species arise.

This kind of speciation is not only natural, it is nature's way of dealing with great changes—climate changes specifically, among others. It's a way of creating new species when new environments appear. In undisturbed ecologies, related species usually evolve barriers to genetic crossing to prevent hybrids from arising. In that situation, hybrids between two species are usually less viable than either parent, and can represent a waste of reproductive energy by the species. Hazels are an excellent example: North America has two native hazel species, both of which grow wild on Badgersett Farm, but no hybrids between them have appeared. Indeed, there is no documented case of a natural hybrid between beaked and American hazels being found anywhere. Attempts to cross them under controlled circumstances always fail; they will not cross. Both, however, will cross with European hazel. The hazels from Europe and the hazels from North America have not been in contact for millions of years, since the continents separated. There is thus no point to barrier mechanisms, and they do not exist.

Many other types of plants are not so fussy. Hybrid oaks abound in nature, and most hybrid hickories and hickory-pecans are found wild, though they are not common.

When great environmental changes happen, species that have been separated for long periods will move into the disturbed places, and meet. And cross, since they will rarely have barriers. Because this happens in a disturbed ecosystem, it is more than usually likely that new genetic combinations may find new places to live, and specialize, and become species of their own.

This was the path to new domesticated crops I specifically set out on in 1984, when I began assembling and planting hundreds of hybrid chestnuts and hybrid hazelnuts. Our chestnut hybrid swarm now consists of five thoroughly mingled species; the hazel swarm, of three species. I also made new collections of wild species, and acquired several more species for reference, and for crossing should there ever be good reason. At this point the hazel swarm is demonstrating all the characteristics we have identified as desirable (though there may be others we don't know about yet). Making additional crosses just for the heck of it does not seem like a sensible idea.

Hybrid Swarm Genetics for Domestication

I have always considered maize, or corn, as the ideal model of domestication. The change from teosinte to maize was so spectacular that the two plants are, by any reasonable measure of biology, no longer the same species. You will still find academics claiming "we have no idea how the Native Americans did it," but that is simply not true. To any trained evolutionist, it has always been obvious.

Hundreds of thousands (probably millions) of people looked very closely at many millions of teosinte/maize individuals over thousands of years, as whole villages participated in harvest. They picked out any genetic variant they thought might be interesting, or better in some way, and planted seed from it apart from the regular crop to see what the next generation would look like. When any new mutation or other type of genetic change showed up, they noticed, and they saved seed from that plant.

FIGURE 12-1. Wild teosinte is on the left, maize on the right, and a hybrid between them in the middle. This is true domestication, from wild plant to human symbiont. Similar changes can be achieved in the neohybrid hazels. Photo from John Doebley, Wikimedia.

They accumulated a huge number of genetic variants and traded them to other villages; modern corn breeders still search their variants for genes we need. The changes piled up. Teosinte normally has two rows of offset seeds. My guess is that the first variant they noticed was plants that had four rows of seeds (this can be a very minor genetic change). Someone saw this, or something equally obvious, and planted it—and the breeding process began.

It took a very long time, but they easily had the time and diversity for the changes that resulted. A major factor in their success was that many cultures made maize a part of their religion, ensuring very careful attention indeed. We can do the same things much faster with modern knowledge and modern tools. But it still cannot happen instantly, or without multiple generations of selection. Or without knowing and tracking what you are doing, where you are in the process, and what your goals are.

As breeders, we are working with genes, and available genetic diversity governs how quickly we can make changes in plants.

P. A. Rutter's Definition of *Gene*

I think defining *gene* has become a form of recreation for geneticists. With every new revelation about the details of genes' interactions, 20 different new definitions appear in the literature, replete with "except in the case of" and "there are also these kinds." The definitions may include fancy terms such as *DNA and RNA messengers* (scientists have identified at least 15 kinds of RNA that all work differently), but I prefer this:

A gene—is one bit of information.

Preserving and communicating that bit, that single piece, of information is the gene's sole function; its form and chemistry vary widely. A gene sends information to the organism and to other genes, so that the information required for organism function is integrated at the proper time.

Versions

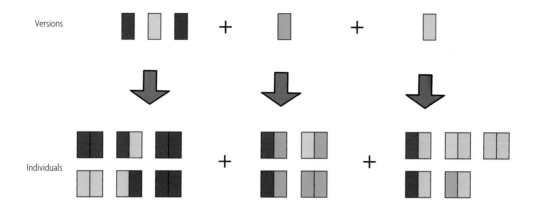

Individuals

FIGURE 12-2. This diagram shows an example of the different combinations of versions possible for a single gene; each version is given a color, and each individual has two of the possible versions. When a gene has three versions (left), six different combinations are possible for an individual. Add one more version of the gene and 4 additional combinations are possible (middle), for a total of 10. Add a fifth version and 5 more possibilities result (right), for a total of 15 overall.

Within a species, many genes will have several versions available in the population. Each individual organism has two copies of each gene, one from each parent. One way of describing the potential genetic diversity available in a population is by using the number of genes, and how many versions there are of each. While an individual organism has only two copies of a gene, there may be many variants of that gene available in the population. Figure 12-2 shows the possible combinations for a single gene if it has three, four, or five versions.

What follows is a very simplified idea of what goes on in the multispecies swarm. Trained geneticists will notice that there are a massive number of details missing in this text; we're acutely aware of them, but just can't fit a fully detailed explanation in this book. If you're interested in the details of why and how to generate and use hybrid swarms in breeding, we consider our Badgersett Woody Ag Short Course to be your best current option. (See *Resources* for more information.) We do intend to publish more broadly on the subject soon.

Given that, let's say we've got a fairly simple trait like hazelnut husk thickness, primarily influenced in Species A by two genes; Gene 1 has three versions and Gene 2 has four. (Refer back to figure 12-2 now.) There are 6 possible combinations for Gene 1 and 10 for Gene 2: $6 \times 10 = 60$ possible combinations of genes for this trait in Species A.

Now let's add Species B to the mix, which has versions of these genes that are mostly the same, but which also has one version for Gene 1 and two versions for Gene 2 that are not present in Species A. Thus Species A and B together have 4 versions for Gene 1, which means 10 possible combinations. They have 6 versions for Gene 2, which means 21 possible combinations: $10 \times 21 = 210$ possible combinations for hazelnut husk thickness in the A–B species swarm.

That's quite a bit more than 60, but there's a much bigger factor that can come into play as well. Let's say we now add a third species—Species C—to the swarm. Species C adds one more version to Gene 1. The kicker, though, is that C also has another, *completely different* gene closely involved in determining husk thickness, and that third gene has four versions of its own. This gives $15 \times 21 \times 10 = 3,150$ possibilities. And 3,000 is a lot more than 60. Genetic variation is the wealth of the crop breeder.

This little example hints at another important bit of the process: Adding more genes to the mix *exponentially* increases the number of possibilities. The real numbers of genes involved in most traits—and the number of versions of each gene—are much larger than the numbers shown in the examples above. And the combination of traits that gives rise to an entire organism is much larger indeed. If we make conservative estimates of the number of different genes and their versions among hazel species, the possible combinations for a few traits can range in the hundreds within one species. Within two species, thousands to millions or more. Within three species, trillions.

If you're talking about a whole organism, the numbers of potential genetic combinations become preposterously, incomprehensibly large when you mix three species. If we call the within-species variation 1, then this makes a two-species swarm about 10^{78} times that, and the three-species swarm is about 10^{122} times the diversity in the single species.

If you've got some idea regarding how very, very large those numbers are, you're probably aware that harnessing the power of this process is going to be a bit of a trick. It's easy to just get lost in all that variation. This is where the details become crucial for ending up with something useful, and the whys and hows of that could take a whole book on their own. Until then, however, here are some important conclusions:

1. Only select for two traits at a time. More than this and it quickly becomes necessary to examine more plants than humanly possible in order to find what you hope for. Eight traits at once and you'll probably have to plant the entire surface of the earth—or, far more likely, you just won't get what you're looking for.

2. The hybrid swarm process uses a widespread natural phenomenon to give rise to combinations not previously existing, and not *possible* with other methods.

3. With this much variability, there can be "bottleneck" plants—individuals that can carry critical traits forward, but will themselves

not be good producers. Discarding "runts" can be a big mistake; knowing what the potentials are is necessary.

4. Backcrossing to pure members of one of the species involved loses the power of the swarm. This process can move a few specific traits into said species, but is then limited to within-single-species variation.

5. The really interesting stuff (the most novel combinations of genetics) *doesn't even start* happening until the second generation following the interspecies cross.

Most natural hybrid swarms are two species. A few are known to include three. As a method of generating new genetic diversity, these swarms are unparalleled.

Eventually, you have to choose the new combination of characters you want, and then restabilize the genetics—decrease available variation so that they breed true. This is part of the reason this breeding path has not been taken before. The amount of data you must collect is huge, and the number of unsatisfactory plants you must identify and discard is high—too high for universities on three-year grant cycles to undertake. The consequence of believing it's "too much, too difficult," however, is: No truly new crops. We'll grow rice and wheat until our soil is gone. What university today would look at wild teosinte and start the work to breed it into maize? Not one.

With neohybrid hazels, we believe the end product will more than justify the time and work needed. The common shortcut to creating a new moneymaking fruit or tree crop is to find three or four "good" plants at an early stage of hybridization, and clone them. Plant thousands of acres with just these three or four genotypes, and care for them following the industrial agricultural model (naked soil, and spray everything). This is why all clonal tree crops go through periodic catastrophe when new pests arrive and move into populations with no resistance and no genetic diversity. Citrus greening has no cure. Oregon hazel fields are still being bulldozed because of EFB. Coffee leaf rust. From 1850 to 1870, France lost 40 percent of its wine orchards to *Phylloxera* aphids from North America.[2] Whole vineyards in New York froze out last winter. Bananas. The list goes on.

Our goal is a neohybrid population that has characteristics such as EFB resistance and cold hardiness fixed. And in fact, we have already achieved fixation of those two traits. The new generations of neohybrid hazels are so consistent for those traits, we no longer bother to collect data on them. We're now working on fixing the next necessary traits, which have to do with crop bearing.

Our breeding processes are working; we can measure the changes in heritability. The end result will be fields of hazels that are uniformly disease-resistant and cold-resistant, and that bear crops heavily and annu-ally, with all of the plants bearing nuts with desirable flavor and size. *But* the bushes will not all be clones, and should be genetically different in many ways from the others.

This is an extremely important difference from the industrial clonal model, because the next time a new pest arrives, or the climate shifts a little, your field of diverse neohybrid hazels will have a good chance to include some individuals that can tolerate the change, from which you can then build a new population of plants—exactly as evolution does in nature. Cloned and genetically identical plants—selected years ago in a different place and climate—simply cannot be expected to provide you with any way to adapt.

The Badgersett Neohybrid Hazel Breeding Cycle

There are advantages to having studied genetics extensively and having been forced to do the horrendous equations sometimes needed in popu-lation genetics. Primarily you learn that you do not have to *guess* how a breeding program will work out—you can translate it into mathematical terms, and thus reliably calculate probable outcomes. Guessing is all too often equal to hoping or just knowing it will work, both of which are very poor breeding strategies and reliably result in decades of work and invest-ment with no success.

Based on the actual known numbers of genes and their available variants for similar characteristics in other species of crops and trees that have been more thoroughly studied, it is possible to come up with reason-able estimates of how many genes are involved in Trait A, and how many gene variants we could reasonably guess might be found in the species of hazels. Some traits are known to be fairly simple; others are known to be typically very complex. When you put numbers to them, you can start to calculate your probabilities in the next generation. We've done that, and we've given you some sense of the enormity of the numbers. Fortunately, you don't have to be able to do such calculations yourself in order to ac-complish some practical breeding work with your hazels. But you do need to become familiar with a few base assumptions that will guide your work.

Start with the assumption that a commercially viable hazel bush should exhibit nine essential traits that will lead to reliably good annual crop production. We can estimate that Trait A normally has roughly three genes involved, and each of those genes for Trait A will likely have, on average, 2.2 possible gene variants. When we do the calculations for all nine traits, we find that their combination necessitates 27 genes; this translates to about 5.7×10^{14} possibilities. If we "lock out" those genes in the overall estimate of the hybrid swarm variation, we can estimate our rate of success . . .

When we crank out those numbers, the result shows that to find one plant with all nine traits assembled after planting out just one generation would require planting the entire surface of the earth in hazels at a 6-foot by 6-foot spacing. That's 10 trillion plants. We suggest not trying to do it that way.

If we select for only two of those nine traits at once, the work becomes tractable; the odds are about 0.002:1. In practical terms, this means that if you plant 5,000 hazels, you could expect 8 individual plants to meet all criteria for those two traits. Add a third simultaneous trait and the odds change to 0.0001:1, which is starting to look next to impossible again. Thus, the best strategy is to select for two traits at a time. Using that strategy, in just five generations of 5,000 plants each, you should be able to do a good job of developing plants that exhibit all nine essential traits.

Now, we do not call our groups of 5,000 (at least) hazels *generations*, because that term can be easily confused with *first cross*, *second cross*, et cetera. Instead, we use the term *selection cycle*. Our first selection cycle contains hazels from six to eight different generation types (F_2s, first backcrosses, first backcross F_2s, et cetera!) What all our starting parents had in common was that they were hybrids, and shared both traits we wanted to fix in our first cycle: resistance to EFB and cold hardiness in USDA Zone 4a. Starting in Cycle 2, we began to push the chances in our favor by removing the male flowers from the hazel plants that were not entirely expressing the traits we were seeking (and yes, over the course of 12 years, we probably removed close to a million catkins). Thus, the assumed "random assortment" was not random but skewed in our favor. We call this semi-controlled crossing, and the results indicate it is quite successful.

A summary of our selection work to date:

Cycle 1: EFB resistance and cold hardiness.
Cycle 2: Heavy nut crop, annual crop production.
>   In Cycle 2 fields: Cycle 1 traits were inherited at about 70 to 80 percent.
Cycle 3: Larger nut size, more standard hazel flavor.
>   In Cycle 3 fields: Cycle 1 traits were inherited at about 80 to 90 percent; Cycle 2 inherited about 40 percent.
Cycle 4: Big bud mite resistance, weevil resistance. (We have planted 3,500 of the 5,000 bushes for this cycle.)
>   In Cycle 4 fields: Cycle 1 traits now inherited at about 90 to 95 percent; Cycle 2 inherited about 50 percent; Cycle 3 data not complete yet.

Cycle 5: Machine harvestable, borer resistance. (We are just starting to plant bushes for this selection cycle.)

No data yet on heritability in Cycle 5.

By way of explanation, Cycle 2 does not end when Cycle 3 plantings begin; we continue to refine the selections in previous cycles, and continue to take data on older cycles to improve statistical samples and to test for new conditions that may occur. For example, Cycle 3 was more than half planted before big bud mites appeared in any of our fields; a few Cycle 2 selections had to be discarded because of catastrophic levels of the mites.

Note that while we say the Cycle 1 traits of cold hardiness and EFB resistance are now fixed in our population, the heritability is not at 100 percent. It will not likely ever reach 100 percent, because there may always be a few combinations of genes that could cause a reversion to a previous behavior. Still, knowing that 90 to 95 percent will be what you need is quite good enough to plant with.

What we are intending to do ultimately is the equivalent of creating a new species—at least to the extent that very few biologists could look at maize and imagine teosinte as the progenitor, or a Chihuahua and imagine the wolf ancestor. We know those are true now, and that genes can indeed be that plastic.

The concept of guiding the evolution of crops and industrial microbes is not unfamiliar to other workers.[3] I'm aware of the actual commercial application, with proprietary details, of very similar step-wise selection processes being used to adapt microbes for biofuel production.

It works.

A Crowd-Sourced Crop

Language changes constantly, and in attempting to communicate these ideas it hit me that the current vernacular for how we envision neohybrid hazels developing in the future is by "crowd sourcing." It's not exactly what the Native Americans did, but it's the closest to it we're likely to come. They had large numbers of people involved, and we need that again. We have demonstrated that progress in a planned direction can be made in the human time scale—but for neohybrids to be as useful as we hope, they must be planted in many more locations, and watched and selected by many more people.

The neohybrid hazels are on the track to domestication.

Please notice that I said "on the track," not "domesticated." Hazelnuts are an internationally traded food, but most present hazel production is what I call old-school coal-fired horticulture, which relies on naked soil monoculture, cloned crop plants with nearly zero genetic diversity, and constant pruning and spraying of pesticides. There are a few surviving

FIGURE 12-3. One future, already here and growing. What will your hazel fields look like? There are plenty of paths yet untried.

bits of ancient traditional silvopasture still found in remote areas. As a food, hazels today are a luxury, not the equivalent of maize or rice. The neohybrids, which can be the future, are a work in progress. Current genetics have the potential to feed you and your family, and produce a modest cash crop; full domestication to feed our cities will happen only when many growers examine many new genetic variants—finding locally adapted versions and the rare new combinations that lead to jumps in production—and learn how to manage them in new ways.

CHAPTER 13

Climate Change, Resilience, and Neohybrid Hazels

This chapter is about how—perhaps—we may adapt to the reality of climate change, and about the many advantages that neohybrid hazels offer over conventional crops in this time of uncertainty.

Our planet, and many of its biota, have been through extensive climate change in the past, but our own species has not ever lived through anything like the broad, rapid shifts now being measured. Yes, there were humans of various kinds during the last ice age, but there were also places to the south that were largely unaffected by that event, where humans lived as they had before. What's coming is different.

Our expected climate patterns have been rapidly upset by humans releasing billions of tons of additional gases (CO_2, methane, and a considerable number of others) into the atmosphere, so it no longer radiates heat back into space, at night, at the normal rate. Unlike previous climate variations, this time the variability is most likely to keep going, growing hotter and increasingly extreme. The gases keep the heat in, and the winds will change, the heat in the oceans will change, and so on.

Most particularly, the technology-dependent species we have become in the last 200 years has never faced anything like what lies before us. I know just enough about science, and climate, and climate change to warn you that anyone who makes certain predictions about our future does not understand the situation.

We do not know what we are facing; nothing like this has happened to us before. If you review recent papers published in respected journals, you will find that scientists keep coming across newly discovered manifestations of climate change that were not on any scientist's radar or worry list as little as two years ago. We absolutely do not know what will happen next. How do we prepare for that?

As I see it, there are three choices. Give up. Pretend it isn't happening. Or try anyway.

For me, as an evolutionary scientist, the choice is obvious. The first two options will lead, quite certainly, to non-survival; your genes will be de-selected from any future. Pessimism and despair are philosophically pointless, I think. And I will bet you anything that trying will be more fun. So here's my motto: "Adapt. Why not?" Sounds flip, but it isn't.

Use your brain. It's our biggest tool. Analyze and dissect any aspect you can think of, and see if you can think of any way that might bring us (and your genes) through what could be a truly tight evolutionary bottleneck for humanity. Whether we will really face "The End Of The World As We Know It" (TEOTWAWKI in Netspeak), we cannot know until we get there. I am not among those who believe it is absolutely inevitable. But I do think it will take great resilience along with some extraordinary—and unforeseen at this point—development to avoid it.

The easiest way to comprehend the details of climate change on a human scale is to look at a boiling pot of water. This is a "thought experiment." Picture putting a 2-quart pot of plain water on the stove to heat. In a few minutes, it will heat up to the point where some water right at the bottom of the pan is momentarily heated to vapor. This causes a boil—steam and water rising rapidly to the surface, then the water rapidly moving back into the mass in the pot. Fun to watch. (Yes, the physics of boiling water is slightly more complex than that, but it doesn't matter.) If you heat it just right, you will achieve a "rolling" boil; when the heat is adequate and stable, the pot will boil in a stable way, at a steady pace.

Got your steady, moderate, rolling boil? Good. Look in the pot, and really watch the roll. Very likely, you will quickly notice some patterns that repeat; the rolling rises are somewhat predictable in when and where and how often they rise. If you can't see this in your head, go boil a pot of water and look! Okay. Got a sort of pattern in the pot?

Now turn the heat on the stove up just one notch.

You know what will happen: The boil will become faster. And your old pattern is gone, usually totally changed. That's your change, both in weather and in climate: winds, currents, seasons, jet stream patterns, all different.

Now in both the slow rolling boil and the fast boil—what is the temperature of the water?

You know, and I know, that water boils at 212°F (100°C), period. (No smartypants quibbles about sea level and 1,013.25 millibars, please; you know what I'm talking about.) Boiling plain water is never 213°F or 101°C—it must turn to water vapor when it's that hot. All the water in the pot is the same temperature, whether it's at a slow boil or a fast one. What has changed is not the overall water temperature, but the rate of heat flux through the water. (Well, and the rate of evaporation, which is okay by me.)

Very similarly, our average air temperatures do not have to change much for the increased heat flux from sun to air, water, land, and space to "boil atmosphere faster." Bigger, faster weather events, shifts in seasonal patterns—everywhere. It's a metaphor, it's not exact. In reality our world is actually getting hotter, but the effects on climate and weather are already much larger and faster than you'd guess by simply looking at average temperatures.

The bubbles in your hotter boiling pot bubble up higher in the pot, and faster. That means what you've already heard many times: We expect more storms, bigger storms, and in general more extreme weather variations, more often. We're already seeing this, of course. Our own county here in Minnesota has been declared an official federal disaster area twice in the past 10 years, for flooding. We've had two "1,000-year" rain events in 10 years.

I've been attending and speaking at conferences on climate change since 1988. And even at the early climate conferences, it was obvious to scholars what many of the possible ramifications of climate change could be—possibilities that have still not really hit world public awareness. Water sources for entire countries will disappear. That causes wars. Starving people from collapsed countries will cross national borders, laws, walls, and guns notwithstanding. Slavery, in all its variations, usually thrives then. Hunger, from chronic undernourishment to famine. Diseases will migrate with people and new weather. Despair and paralysis of will. All those things are already happening. They're almost certainly going to get worse, and closer. If you believe in gravity. And physics.

In the face of despair, I say: "Adapt. Why not?" Humans have a bone-deep primal need to be useful—to our families, our villages. Doing nothing of value kills us, fast or slow. I contend that working to find a way forward is useful. And will be satisfying.

Humans have, historically, dealt with all these problems before: war, drought, disease, hunger, even climate change. There are things about resilience we can learn from history and put to use as we find our way forward. As an easy example, humans who are loners tend not to survive in such circumstances. Communities may.

12 Advantages of Neohybrid Hazels

And with that, let's turn our attention to neohybrid hazels. Do hazels, and/or their hybrids, have characteristics that might help them survive in potentially semi-chaotic weather and climate? In fact, they do, and not entirely by accident. Although I originally conceived my quest for a new food crop as an answer to agricultural problems such as erosion, water degradation, and loss of biodiversity, I looked for species that were tough under multiple threats, too. It turns out some of the resilient characteristics that hazels evolved were to cope with disturbed climates.

Advantage #1: A Long, Long Track Record

Hazelnuts have been identified in the fossil record at least 40 million years back. Beyond a shadow of a doubt, they have survived many severe climate shifts—and they're still here, still hazels. And where were they living all that long time? Were they lazing in some rain forest in the tropics? For the three species in our current hybrid swarm, definitely not.

Based on the fossil pollen record, we know hazels were often among the very first woody plants to occupy land recently vacated by continental glaciers.[1] Disturbed land. With a harsh and changing climate. They don't have to have disturbance to thrive, but they thrive in disturbance. Sounds hopeful.

According to our own observations, much of this ability to tolerate changes in microclimate can be a matter of genetic plasticity, not different genotypes. Twenty-five years ago, we planted a row of hazels in full sun, and recorded their performance. At the same time, we planted a row of hickory-pecans just to the north of the hazels, and a row of chestnut trees just to the south, so the hazel row would become heavily shaded. The same hazel plants changed their growth form, from spherical sun-adapted dense bush to a taller, sparse, layered understory form. One genotype can change the way it grows to thrive in either sun or shade. Sounds very hopeful.

Another related aspect here: We have reason to think, based on measurements of other bush species, that hazel bushes may easily reach an age of around 1,000 years (I would so love to actually measure that someday).[2] We won't attempt to claim the 11,000 years measured for creosote bushes, but a measly 1,000—almost for sure. No, not the tops; individual stems may live only 8 years—up to 20, depending. The roots, and root crown, certainly live hundreds of years in nature, though, and probably far more.

Even in a 500-year life span, a hazel propagated by seed at the beginning of that time must be able to reproduce successfully in a changed climate, in the middle or end of its life. Or the genus would not still be here, after 40 million years. One plant has to carry the capacity to adapt to large change. Hopeful, again.

Advantage #2: Lots and Lots of Genes

Enormous genetic variation—unprecedented in agriculture—is in the neohybrid gene pool. What that means is that in the Lottery of Survival, the neohybrids will provide you with about 10,000 lottery tickets, instead of the single one you'd get from traditional crops bred for the climate 50 years ago. That's intended to give you a feel for the difference. The real difference is actually a lot bigger than 10,000 to 1, but larger numbers start to become meaningless to human brains.

We find it consistently difficult for geneticists not deeply familiar with population genetics and mixing multiple species to believe, or even comprehend, the genetic diversity of the neohybrids. We've checked our math on this (and we'd be delighted to send our calculations to you if you email and ask). Our math is quite conservative regarding the assumptions about, for example, the number of gene varieties available. (Note to professionals: Yes, by "gene variety" we mean "allele"—but if you've ever taught genetics, you know what a problem that word is to teach or understand.) There is a gem of a research paper[3] from the University of California at Berkeley that reports measurements of genetic diversity in the remnant populations of California hazel (considered a subspecies, at the moment, of beaked hazel). By measurement, the researchers found some small populations where the number of varieties for one gene was 21. By comparison, for our calculations of hazel swarm diversity, we made the assumption of an average of 4.2 varieties per gene. If you're familiar with the math of combinations and permutations, you'll know immediately that with 21 potential varieties instead of 4.2 varieties per gene (and from three species, not two), the potential diversity explodes. The difference between 3×4 and 3×21 gives a glimpse of it, though the mathematical reality is much larger than that. It's the difference between "That's pretty okay" and "You're putting me on." But we're not putting you on.

The neohybrid hazels include genes from warm Mediterranean populations, temperate regions, and the extreme northern temperate regions (nearly subarctic). If that diversity is maintained, it is reasonable to hope we can find variants that will adapt to many varying conditions. The genus also includes other species we have not incorporated, from much lower latitudes in India and China. Crosses have been made, experimentally, but no advanced generations have been developed. We might need those genes too, someday—if they are still available.

Something rarely discussed in conversations about needing to move Food production north, away from increasing heat, is that there will be a major problem for existing crops if human populations must shift to currently colder—more northern—latitudes. The annual sunlight distribution at latitude 40 degrees north (just about Indianapolis; mid–Corn Belt) is what our agriculture is used to. But at 50 degrees, you're in Winnipeg. It's not a matter of when it thaws, and when it freezes—that's going to change. It's a matter of the angle of sun. The days farther north get much longer than crops are used to in summer, and their nights much shorter.[4] The crops will need to be able to cope with the new circumstances. Corn is from Guatemala, and has no genes for high latitudes. Soybeans are from South China. Hazels go all the way up to 51 degrees, maybe 53 if we explore—and they bear nuts there. Where do you want to place your bet?

We have reason to think that the latitude adaptations in hazels are genetic. We've collected specimens of both *Corylus americana* and *C. cornuta* above 50 degrees north latitude, and after twenty-some years growing here in southern Minnesota, they all have fared pretty poorly. They are far less robust than their local cousins. We've attempted to give them resources, such as weed control and fertilizer, but nothing has ever worked. Thus we think it's the light differences that are telling. This isn't scientific proof yet that it's the latitude shift that is the problem, but studies done on European hazels and other species show that this does happen.

Measurements of "whole-crop carbon," where the entire plant, including roots, stems, leaves, seeds—everything produced in the year—is captured, dried, and the carbon content calculated, indicate that a field growing one crop of maize in a year will capture only about a third of the carbon that the same field growing hybrid poplar would.[5] So far as I am aware, I was the first one to compare maize to poplar. Agronomists and foresters almost never talk to each other, so I had to do a little cross-pollination. Hybrid poplar was for some time the "white rat" experimental organism of forestry, with a great deal of quality research achieved before the grant fads moved on. Three times the energy is being captured by the trees. Potentially, we could skim a third of the trees' energy as a food crop, and still have twice the energy captured by maize, including the stover and roots, allocated for the trees themselves. And not at all incidentally, such fields would be capturing three times the carbon now captured, much of which would be sequestered long-term in the wood and roots. Much of the trees' allocation could eventually be sequestered for thousands of years, if converted to biochar.[6]

An overlooked but seminal publication was brought to my attention by one of its authors, Dr. Don Lawrence.[7] In the constant battle to measure and compare the "productivity" of differing ecosystems, rather than attempting to measure biomass produced per year (extremely tricky to do accurately—think of measuring annual production in a cattail marsh, for example . . .), he and his student hit on the idea of measuring the chlorophyll actually present in various ecosystems over the course of a year. Rather than directly weighing biomass—some of which will have volatilized and blown away (like terpenes from pines, or the sap that many tree species drip on your parked car); or blown away before being measured, like pollens; or been eaten by insects below the soil or up in the canopy—the weight of chlorophyll per acre gives an estimate of how much work the plants are capable of doing, right now. And while it is true that very old leaves do not appear to photosynthesize as rapidly as new leaves, it is also true that in all plants, if green parts are covered and kept entirely dark for a while, they will turn white. Plants resorb and recycle the materials of chloroplasts if they are doing no work. All plants. Likewise, if you put those now white leaves or stems back in the sun they will turn green again as chloroplasts multiply in the existing cells—because there is now work

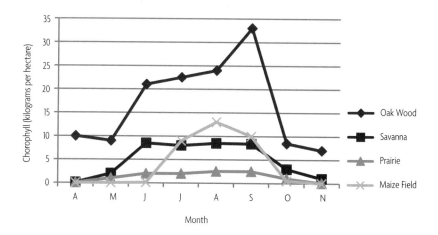

FIGURE 13-1. Graph after Ovington and Lawrence, 1967. There is vastly more chlorophyll present in a forest than in a prairie, a savanna, or a field of maize, and for a far greater portion of the year. This is entirely consistent with the measurement of hybrid poplar producing approximately three times the biomass that maize does.

for them to do. It is an excellent measure of potential work done, one that deserves more use in attempting to quantify plant productivity under diverse conditions.

While we're talking about genetic adaptation to climate change, there is also a common contention that C_4 crops are a better bet for our future than the usually C_3[8] wild ecosystems. We believe this requires a great deal more examination, and is actually contradicted by peer-reviewed research already published.

The Gates Foundation is one of many organizations now funding an effort to transform rice to a C_4 plant.[9] There are many caveats, however, one of them being that the C_4 chemistry is only efficient at high temperatures; 82 to 95°F (28 to 35°C) is a frequently cited range. Below 82°F, C_4 efficiency drops rapidly; C_3 efficiencies are then more stable and adaptable.[10]

Maize, sugarcane, sorghum, millet, and amaranth are C_4 food crops; all other crops are C_3. In nature, all C_4 plants are from warm climates, none from temperate or cool. Naked soil is hot. Shaded leaves are much cooler.

One of the potential fallacies of the C_4–C_3 comparisons is that the C_4 advocates are assuming an equal amount of leaf area, or chlorophyll, for the C_4 and the C_3 crops. With any woody crop, that is an inappropriate assumption. For the neohybrid hazels in particular, their mature height is around twice that of corn. They are also photosynthetic for a much greater portion of the year: fully leafed out and at work as much as a month before the first new corn shoots appear on the naked soil. When sunlight is at its most intense at the summer solstice, the midpoint of the year, it is common in the US Corn Belt for sunlight still to be falling mostly on bare soil for row crops, while it's dark and cool under nearby trees. And woody plants are still photosynthesizing, pulling CO_2 out of the air a month or more after this year's maize is dead and drying down.

Advantage #3: Superior "Coppice-Ability"

Coppice is a fun word. Lots of folks in the United States have no idea what it means; in the UK everyone knows it—but they pronounce it, and often spell it, as *copse*. There are regional differences in details of the meaning—and the usual regional assertions that their definition is the right one.

In the UK, it is usually a noun and refers to a group of trees or bushes that has been traditionally cut to the ground, wood harvested for various uses, and allowed to grow back from the ground, often for hundreds of years. Sometimes thousands.

In North America, *coppice* is usually a verb, meaning "to cut down, then regrow."

Hazels are exceptionally good at coppicing, and we select neohybrids—those that aren't are removed from breeding—for this trait. There are reasons why hazels are so good at this. For one thing, all three of our starting species have periodic wildfire as part of their evolutionary past. And across Europe and the Middle East, hazelwood has been coppiced by humans regularly for at least 20,000 years; I believe this number will eventually be determined as more than 50,000. It can be fairly stated that for 6,000 years (at least), up to a mere 500 years ago, all European cultures depended heavily on coppiced hazel and could not have progressed without it. The shoots that develop from coppiced plants tend to be uniform in diameter, straight, and develop branches at much wider intervals—all factors making them much easier for humans to use.

At the very earliest edges of history, houses in Europe were largely made of wattle-and-daub construction; archaeology shows it was widespread in the Neolithic, 10,000 years ago, when tools were almost entirely made of stone. The wattle was hazel rods, from coppiced copses. Thatched roofs? Grasses and reeds held in place with coppiced hazel rods underneath. Sheep fence? *Hurdle* originally referred to modular sheep fence constructed of split coppiced hazel. Fish weirs? Coppiced hazel rods—going all the way back to the Mesolithic in Denmark, where research determined they were cut from coppiced hazel using polished flint axes as much as 9,500 years ago (so far).[11]

Bundles of coppiced hazelwood were used to pay taxes in England; bundles were standardized, and undersized bundles were punished.[12] Billhooks, the tools used to manage coppiced wood, are abundant among the tools found in Iron Age archaeology, and are found even into the Bronze

Age. In North America, both our hazel species were used for baskets and even arrows, as well as fish traps.[13] Native Americans sometimes burned areas as a method of coppicing hazel, and found that it increased the number of bushes, along with nut production.[14]

What does all this have to do with resilience? For one thing, in the event of broken supply chains for modern structural materials, all those past uses of hazels can be revived. And if your neighbors need any of those things, you will have a commodity you can sell or trade. Live hazels can be trained into very durable fence, too—hedges "horse high, hog tight, and bull strong." If some climate-related disaster, from fire to a direct hit by a tornado, knocks down your hazels, they will grow back very well.

Here is an unhappy advantage of hazel coppice-ability should we hit TEOTWAWKI. In the event of war, local or larger, smashing, cutting, and burning a hazel field will *not* destroy the hazel bushes; they will grow back. The bushes are not absolutely indestructible, but it would take a highly skilled and heavy application of herbicide to actually kill them. Not impossible—but slow and very costly, and very risky for the enemy. Physically damaged bushes will begin to bear a crop again a few years later. And while you wait until the nut crop returns, you can grow other crops or graze some livestock between the rows of short bushes, as you did when you were first establishing your planting.

Advantage #4: Storm-Proof

In climate change, we have to expect more frequent, and more destructive, storms.

Hazels will survive many very heavy winds, flooding events, and hailstorms. We've done that, seen it. In late May 2011, we were actually hit by a tornado. The bushes were heavy with leaves, but the nuts were very small. We lost some apple trees, and several large red oaks snapped off about 20 feet up—but there was no damage in the hazels. We have had a report from a grower in Iowa that his hazels were hit by a "downburst," with straight-line winds over 90 miles an hour, when the nuts were full-sized and heavy: "The hazels were blown down flat on the ground during the wind. When the storm passed; the hazel bushes got right back up. But the nuts didn't."[15] We're not claiming they are invulnerable or immortal—just much, much tougher than annuals.

We've had hazels come through hailstorms that pounded nearby annual crops into the mud. A few leaves were torn off or ripped up, but the loss of nuts was almost unnoticeable. The deep structure of the bush allows the upper leaves and branches to effectively shelter the lower ones. The flexibility and vertical alignment of the stems likely plays a part also.

We planted hazels in an old pond bottom here so we could observe their performance in very wet conditions. During spring thaw and during local flooding events, the pond still fills with 3 feet of water. The hazels tolerate being partly underwater for a week, at least, with no damage at all to the parts, and crop, above the water. We've heard stories of "under 3 feet for three months." Judging from wild hazels living on floodplains, that may be believable.

Soil erosion is nonexistent in a hazel field, even during 6-inch downpours. There is grass/herbaceous cover between hazel rows, and the hazel roots themselves cling very tightly to soil particles. And because there is no annual tillage, the fields are full of very effective "bathtub drains"—mouse and other rodent holes, which carry a tremendous volume of water quickly belowground, where it is absorbed into the soil without erosion.

Advantage #5: Very Low Inputs

Neohybrid hazels require far lower fuel inputs than crops that need tillage every year. And no pesticides. Both of those commodities are likely to become much harder to get, and much more expensive when they are available. And remember that you need to burn fuel to apply pesticides on large fields, too. Hazels do require some fertilizer for maximum crops and growth, but you can use animal wastes for this—and if nothing is available, it doesn't mean no crop, just less that year.

If you grow hazels on a large scale and harvest by machine, you will need some fuel to power the harvester. Remember that hazel oil can make excellent biodiesel; we've even burned it raw in our diesel engines. Not that we're superstitious or anything, but when we acquire something with a diesel engine, we "baptize" it with raw hazel oil in the fuel tank. Just a pint or so.

It's worth pointing out (again) that *traditional* hazels do require annual soil tillage, to prepare flat naked soil for harvest, and do need multiple applications of pesticides and anti-sprouting plant hormone, nearly every month of the year.[16]

Advantage #6: A Diverse Supply of Products

The neohybrid hazel crop yields an exceptional diversity of products, with an exceptional range of utility. Your hazel plants can directly provide you with fuel, both wood and liquid; food for your family or animals; construction materials from which you could build a house or barn; material to build fences; materials easy to sell or trade; and more. And don't forget the variety of things that can be grown between or within hazel rows.

The soybean industry prides itself on the huge array of products soy components can be found in, and rightly so. But virtually all of those rely on industrial separation processes, and complex syntheses using many other ingredients. Even to be used as animal feed, regular soybeans must be heated to make them non-toxic. (The reason some soybeans are called "edible soybeans" is that most soybeans aren't.) All those soy products simply cannot be made by the individual grower, or even in a small community.

Advantage #7: Weathering Bad Weather

Bad weather this year? Hazels can make a crop anyway, using resources stored in previous years. Our database has shown us for multiple years that, unlike many fruits that depend directly on the current year's photosynthesis, hazels can pull stored resources from previous years to ripen a crop despite, for example, too many cloudy, rainy, or chilly days during the normal growing season. We've seen them do it—actually ripening the nut crop two weeks early in a cold year, for instance. The bushes cannot keep this up forever, of course; eventually stored resources can be depleted if they aren't replaced. Our database suggests that three years may be the limit before the crop crashes. That is a far greater margin of safety than any annual crop can offer. Dr. Greg Miller has done very meticulous research showing that in chestnuts, fully 50 percent of all the dry material in a ripe nut is pumped into the nut in the last two weeks before it drops.[17] That's approximately the same as if apples and peaches doubled their weight in the last two weeks (they don't). Fruits, like apples, are almost entirely sugar, cellulose, and water, and are the products of this year's sunlight and rain. Nuts are mostly oils, protein, carbohydrates, and minerals—far more complex and nutritious. Dr. Miller's chestnut research and our repeated observations on hazel ripening times, and their ability to bear for years without fertilizer (not forever!), together lead to one conclusion: Nut trees do *not* rely on this year's sunlight to make their crop, but rather mobilize and translocate resources stored in previous years if necessary. We think hazels may mobilize minerals and energy stored at least three to four years ago.

Imagine what that could mean to the world food supply. Cold, with no sun this year? The nuts will bear their crop anyway. Annual crops, by their nature, will never be able to do this.

Advantage #8: Local Climate Modification and Moderation

Compared with a field of corn, a field full of densely planted hazel bushes, most over 10 feet tall, will increase local humidity, reduce surface evaporation, cut wind speeds, moderate temperatures, increase water infiltration and retention, slow snowmelt, and provide habitat for thousands of species, making your entire landscape more resilient, not just the plants themselves. If your neighbors start growing hazels, too, the advantages will spread.

Advantage #9: Non-Local Climate Modification and Moderation

Neohybrid hazels capture more carbon than row crops—two to three times as much, we believe—and much of that is sequestered (kept out of

Good Carbon, Bad Carbon

There is currently some confusion between climate scientists and the public about what kinds of carbon are good, or okay, and which are bad. Particularly since scientists tend to assume everyone understands their jargon (like *current budget*).

"Carbon is carbon, isn't it? If we put it in the air, that's bad!" is something you will hear during discussions on biomass fuels.

This bit of logic actually isn't correct, but many people are lost in the long explanations why. Here's the short version: *If* the wood you burn is replacing fossil fuel carbon (coal, natural gas, oil) that would be burned otherwise, *and if* the land where that wood was cut is used to grow more wood, *then* wood biomass fuel is a help to the atmosphere.

A slightly longer version, which I find useful: Before the Industrial Revolution, the atmosphere carbon budget was balanced. Every year, logs rotted, food was eaten; but also every year, new trees grew new wood, new crops were planted. From year to year, then, the total carbon in the atmosphere varied almost not at all. This is the biological carbon cycle. Then we started to burn fossil fuels, taking deeply buried ancient carbon (hundreds of millions of years old) and putting it into the air. Fossil carbon is not counted as part of the biological carbon cycle; it is counted as part of the geological carbon cycle, which involves carbon that moves not every year but only on the time scales at which mountain ranges form and continents drift.

The plants of the world are unable to absorb all the carbon we now add from fossil fuels. Burning even more of them is a very, very bad idea. If you can use wood instead, from trees that pulled that carbon out of the air in the past few years, and then grow more trees to capture that carbon again—the net carbon gain by the atmosphere over, say, 20 years, is zero. This is a huge benefit.

the air) from the global carbon cycle for much longer periods than row crops can provide. In general, carbon captured by maize is returned to the atmosphere in an average of 100 days. That's the "active fraction"; carbon deeper in the soil can be there for decades.[18] And when the cornfield is tilled, the deep soil carbon brought to the surface will be back in the atmosphere next year, not 40 years from now.

Hazel roots penetrate the soil much deeper than corn, and much of the carbon in deep roots is sequestered for a long time. If you burn hazel wood or nutshell to keep warm, or generate electricity, or cook food, that "current budget" carbon will replace fossil fuel carbon—and that helps. And as hundreds of people will tell you these days, if you can make biochar and put it deep in the soil on a large scale, it can stay out of the atmosphere for thousands of years.[19]

Advantage #10: Local Energy Production

Larger-scale hazel crop production has the potential for being a more resilient local energy source. Wood and nutshell can be burned/gasified for electricity and heat; oil from non-food grade nuts can be available for biodiesel—or you can use food grade crop for biodiesel if the need is greater. Much better than cutting wild forest, in many ways.

Advantage #11: Flexible Scale of Production

Neohybrid hazels are amenable to both large- and small-scale production. You can harvest them by hand or by machine. Having options—for the farmer and the community—always provides greater system resilience.

Advantage #12: Hope

We tend to die, without it. Or drop into horrific societies.

Maybe neohybrid hazels can help us get through the bottleneck ahead. And maybe this can be the pattern for a genuinely sustainable agriculture if we find ourselves on the other side of the bottleneck.

CHAPTER 14

The State of the Crop

If you come across an article all about the "new hazelnut crops," and the author writes about "the state of the industry"—watch out. There is no new hazelnut industry yet. It will take at least 10 more years for an industry to develop here in the Upper Midwest, and longer than that in areas where no one has started growing them yet. I'd prefer to see growers in one locality producing, processing, and selling at least 40 tons a year, reliably, and young plantings coming into production before I would consider it an "industry."

One problem with crowd sourcing a project such as creating a new crop is that some people who choose to join the crowd will have lots of enthusiasm, but won't be willing to realistically consider whether they can be successful. They may not live in a place where hazels will grow well, or they may not be willing to spend the money to buy good-quality planting stock, or they may not want to bother fertilizing the bushes. There will also be a few Great American Individualists who will launch out on their own, but end up heading backward because they will insist on following their own untried methods rather than practices that are proven effective. Worse yet, they might decide not only to grow hazels, but to try to produce and sell them to you—with zero understanding of how to improve their hazel seedlings. Unfortunately we know of several who are selling hazel hybrids we consider useless. Time spent learning to distinguish between careful science and no science will mean the difference between a hazel field that can produce and move forward, and one that will always disappoint.

We have sold neohybrid hazels to growers in many locations; our ability to follow up on them is limited, though, and not everyone tends to provide feedback. Figure 14-1 summarizes the data we have collected up to 2012.

The great majority of these plantings are less than 15 years old, and due to consistent difficulties in getting people to fertilize them—at all, let alone enough—not many of these are approaching serious production yet. In phone surveys of Badgersett customers performed by the University of Minnesota in 1998,[1] one of the questions asked was, "Have you fertilized them?" Only about 20 percent said yes, despite clear instructions to fertilize. These are plants selected to produce. If

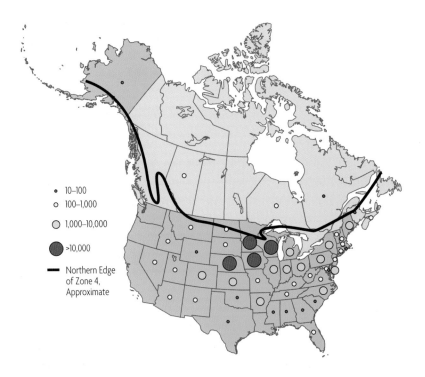

FIGURE 14-1. We keep track of where we sell Badgersett planting stock, which is a reasonably good estimate of where the trees are being planted. This map shows the wide distribution of those plantings through early 2012.

you don't fertilize them, they not only will not produce but will probably kill themselves trying to make big crops with no resources. (See the fertilizing instructions in chapters 5 and 7.)

Have you noticed any large surge of "Locally Grown Hazelnuts!" in your area? Me neither.

At the time of this writing, Badgersett has two plantings: the basic research collection in Minnesota, and a planting intended for both breeding and production in Illinois, from Cycle 2 and 3 genetics. In the research collection, there are a lot of holes in the rows, because we usually do not replant after removing inadequate plants. If we did replant, we'd be able to harvest a lot more nuts, but seeing trends and gathering data would become enormously more difficult. We harvest much of the Minnesota crop for seed, identity preserved, and in recent years we have had to simply not pick some of the nuts because labor was not available. I estimate that we left about 3,000 pounds of nuts unharvested in 2012, and 2,000 pounds in 2013—uncertain but very significant fractions of available production (perhaps half in 2012).

Many rows in the Illinois planting are solidly populated with plants, and nut production has been high. But this planting is where we've been learning how to machine-pick. What we've discovered is that not all of the bushes are machine-friendly. Also, the timing of the harvest has been a learning experience. The first two years our harvesting efforts were quite ill timed; 2013 was much better. We estimate the machine can pick only about half the bushes at this point. And we're using much of that seed to produce our planting stock for Cycle 4, which focuses on machinability. The amount of harvest we have available as nuts for consumption is not great.

Some of the people who grow hazelnuts or want to grow hazelnuts are enthusiastic mechanics. Every time I have explained that someday we'll need a mechanical husker or nut cracker to support a hazelnut industry, some of these enthusiasts build one. To my knowledge we now have about ten neohybrid hazel huskers, all of different design, and at least five hazel-cracking machines. We build them, too, and in terms of the progress of the crop, all this machine diversity is great. We just need more nuts to put through them.

A couple of years ago, a very interested potential grower followed me around closely at Badgersett's annual field day (the third Saturday in August every year), asking every question he could think of. I gave him the best true answers I could. Near the end of the day he turned to me and said, "You're trying to talk me *out* of this."

No, I wasn't. No, I'm not. I said, "No, I'm just trying really hard to tell you the absolute truth."

Look at it this way: Yes, I believe in the promise of this crop. It's what I'm doing for the rest of my life. The potential benefits for my children and yours are very large. But we've got a long way to go before anyone thinks of neohybrid hazels and corn and beans all as commodity crops.

The dangers of assuming "This is good enough!" and running as fast as you can are serious, and were recently illustrated in the realm of hybrid hazelnuts. If you look in the literature 15 years back, you will find quite a few very enthusiastic articles about the "new Michigan hazelnuts!" Besides the popular press, some universities were bitten by the new Michigan bug, and called them "a promising new crop!" These plants were grafted clones, though, and the hype about them was premature. The breeder was an enthusiast, not a trained scientist, and did not appreciate the rigor of testing needed. As it turned out, some large fields of these clonal plants were installed, but Eastern filbert blight (EFB) killed them in spite of the breeder's claims they'd been thoroughly tested. The nursery involved seems to be out of business now. Some of those hazel bushes are surviving in New Zealand—they do not have EFB there. Yet.

That has been the repeated experience with short-term selection of simple hazel hybrids—you can produce plants that bear very nice nuts. But inevitably, the result is: "The plants are dead now; but the nuts were great!" We're trying hard not to follow that path, and our record of bushes that have survived 30 years in an area surrounded by EFB does prove we're on the right track. We think it makes sense for you to try to stay on that track, too—not get off it.

If I were to compare the state of neohybrid hazels today to the development of maize from teosinte, I'd say the most advanced genetics are equivalent to the "podcorn" stage: a cob that has 10 rows of seeds instead of 2, that grows about 3 inches long, and that has a husk around every seed. That's far from teosinte, but it's also very far from modern maize.

 ## Why and How "Real Farmers" Will Change to Neohybrid Hazels

We are writing this book attempting to address the needs and questions of growers at all scales of production, from backyard to 10,000 acres (someday). All of those scales are necessary, and we expect it will turn out that all scales are useful, far into the future, for differing circumstances.

In order for neohybrids to have significant impacts on food systems and the health of the planet, though, some large-scale commodity crop farmers will need to find ways and reasons to grow neohybrid hazels. They have most of the land, by far. Yes, growing 25 to 100 acres of hazels for your family farm can be beneficial for your land, water, and pocketbook—but think how much more it could mean to the world to replace a significant portion of the land now planted in corn and soybeans with the plowless perennial system.

We're thoroughly aware that one common reaction to this suggestion is sheer disbelief: "There's no way regular farmers will *ever* do this." Frequently that reaction comes from folks who are truly experts in those crops. Here's the thing, though: All of history is against them, and on our side. A great many farm folk still remember two great revolutions in our agriculture firsthand: the development of hybrid corn, and the explosion of the soybean crop. And a great many remember hearing our parents and grandparents talk about two other astonishing revolutions in their lifetimes: the shift from horses to tractors and from hand labor to machines. My own father grew up on a truck farm in northeastern Ohio, with nine siblings and horses.

History tells us so much, when you dig into it. And you do need to dig, because while there are courses and textbooks on "the history of crop development" and other aspects of agriculture, all history is prone to being skewed by agendas and viewpoints—and just plain bad information. Historians will tell you that. You need to read multiple sources, and whenever possible, read original statements and documents. I've done that. I'm not going to bury you in references here, but you can confirm these generalizations yourself, if you want to.

Farmers—the overwhelming majority—are good people. They want to do the best they can by the land, to take good care of it—and preferably pass the land and the farming life on to their children. (That description can fall apart when a corporation owns the land, and the farm operator becomes a mere employee. This is a model already wreaking havoc in livestock production, and now spreading into plant crops.) Farmers are also smart. They have to be smart business folk to stay in business these days. Because farming is a *business*, as anyone with fantasies of the good farm life quickly discovers. Farmers are also smart about their own crops; they are highly skilled experts in what they grow, though they may not realize it.

I have a line I use when I'm talking to an audience of farmers and they start to wander off on me as I get into details. "I can see what some of you are thinking!" I'll say, in a break from the talk. "You're thinking—*Man, this is all way too complicated*. You're thinking it's beyond you. It *isn't* more complicated than the crops you grow now. It's just *different*." I expect to get broad audience reaction at that point, head shakes and grins indicating *Boy, are you full of it*. That allows me to grin back and say, "You guys are huge experts at what you do—and don't know it." More head shakes, they're gunning for me now. "Okay, tell me this: Could you take a kid—out of the middle of Chicago—and teach him to grow commercial corn, in one hour? Two?" That starts them running through in their minds all the incredibly minute detail needed to raise corn—and it's enormous. I get a good wry laugh, and they start to see and listen.

Farmers have a reputation for being conservative about change, and many do indeed fit that description. But farmers by their very nature are also risk takers; just ask them. Every year is a gamble on weather, markets, banks, and luck. And they know it. Underneath, they're proud of it. Part of that risk-taking tradition has always meant there are some farmers who love to experiment with new ways. And new crops. All those agricultural revolutions we mentioned above? Came from farmers.

Once again, the most accessible argument comes from the soybean industry. At this writing, the USDA official projection for the 2014 US soybean crop stands at a record-high 3.816 billion bushels from 84.8 million acres planted, with an expected average yield of 45.4 bushels per acre. Those numbers should stagger you.

But how did this tremendous success come about? These numbers will also stagger you. The American Soybean Association was founded in 1919. About 1,000 people came to the organizational meeting. In 1919, there were already 113,000 acres of soybeans in production. A thousand people interested enough to come to an initial meeting is a very substantial number. Likewise, 113,000 acres is a huge commitment.

But here are the two most telling facts: The yield per acre in 1919 was 9.6 bushels. And the crop was picked by hand, and/or with horse equipment.[2]

This was not a successful established crop; it was tiny, with terribly low yields and huge labor required. And incredibly limited markets. But these farmers (not hobbyists, not professors, but farmers) could *see* the promise. They believed in it and invested in it with no guarantees—just their own brains, and the drive to find a better way to make their farms pay a good wage for their families.

That is what farmers actually do. And they will do it again when they see that they *can* grow neohybrid hazels, using machinery they understand, drying them in the grain dryers they already own, selling the crop down at the elevator. And when they realize they can make better money growing hazels than they can with soybeans or corn. Not very far down the road, hazels will be cheaper to grow, per ton, than soybeans ever can be, because of the greatly reduced inputs: no planting, no plowing, no spraying. Farmers will be proud of the environmental benefits, too.

The first soybean growers were not the farmers who were having a hard time making a living; they were the farmers who had enough financial margin that they could afford to experiment a bit. And those are the folks who are already starting to grow neohybrid hazels. How can soybeans possibly compete?

Tools You Need

You think you might want to grow neohybrids?

That would be terrific. Welcome. If you're seriously intending to butt heads with climate change, though, here's some advice. This advice is not specific to hazels, but it applies to them, too.

You Need Science

Science is not all that hard to learn; it's more that it's often taught badly, which turns people off. Scientists write science textbooks not for students but for state boards of education, which choose them, love to have "all the latest" in them, and have no idea the books are incomprehensible. I've been part of that process. Good scientists will try to sneak some interesting education past the boards, but it's an uphill climb.

Science is an algorithm, a set of rules that—when you follow them honestly—will present you with facts. Facts are incredibly useful; they work, each time. "Knowing" something is true when it actually isn't will get you killed, especially in an uncertain world.

One reason people turn away from science is that they've acquired the common misconception that science is supposed to provide wisdom, or a sort of abstract truth. That's simply not the case. It doesn't and wasn't ever intended to. Finding wisdom is your responsibility.

In practical terms, when you're growing hazels, or pursuing any venture, scientifically, take lots of notes. *Always* keep a "control" planting when you change something or you will never learn a thing. Ask others what you've missed.

You Need History

You can learn a lot from the failures and the stunning successes we have already experienced with trees—stories that are unknown to most. There are an abundance of failures to learn about, like the extensive hazel breeding program at Cornell in the early part of the 20th century. The goal of this program was to breed cold-hardy, EFB-resistant hazels—with big nuts. But the program was shut down because they never made progress. We do have some of the surviving genetics from that program; the mortality and reject rate is still high.

And breadfruit. Most cultivated breadfruit are species hybrids. Most can no longer reproduce without humans. The Pacific Islanders kept dozens of specific cultivars with specific growing and cropping behaviors to suit different needs.[3] Spectacular plant breeding by another primal people.

On the downside, about once a year another Westerner (European culture derivative) "discovers" breadfruit and announces that it will solve all world hunger problems in the near future. Without bothering to find out that the crop has highly specific needs, and is not adapted to very many places. That's been going on for about 300 years now, without much improvement in Westerners' analyses or memories.

History will teach you what has not worked, and what has, and will give you lifesaving hints.

You Need Genetics

You don't need to know everything about hazel genetics, just about as much as a dairy farmer needs to know about his animals. Enough so that as you study your hazels and how they are growing and producing, you can think also about their parentage, and make a good guess as to what characters you can try to add next to your own hazels, in a realistic fashion. Just planting a lot of hazels will help, certainly—but it will all work far better and faster if you have an educated eye to see with.

You Need Communication

Communication is an art. Can you tell people what you've learned? Can you listen? As with all arts, practice will make you better. The most succinct statement of the problem I know of is from George Bernard Shaw:

The single biggest problem in communication is the illusion that
it has taken place.

I can't tell you how many times I've found myself on one side or the other
of that exact situation.

You Need Community

Both a neohybrid hazelnut community, whose members help one another
learn and succeed, and the other kind: good friends and neighbors.

There's one thing that all the preppers, all the doom bloggers, all the
sane people starting and moving to Transition Towns all over the world
have come to agree on in the past few years: If there is a crash like the kind
we fear, no one is going to survive it alone.

It'll be plenty hard enough with a community trying to make it.

All of This Needs—You. There Is Much to Do.

If you thought you were going to pick up this book and find "All the
Answers! Ready to Go! The New Miracle Crop!"—I'm sorry if you feel
disappointed.

There's a lot of work ahead. And that's true whether the subject is
hazels or anything else. Brandon and Sue and I are going this road with
neohybrid hazels. Caveats and all, it looks more hopeful than any other
we've found. We've made progress already. What's needed now is more
people to join in and grow this crop—exactly like the many communities
that worked together to produce maize from teosinte. We do think you're
likely to do better at this crop if you approach it carefully, scientifically,
with your sleeves rolled up and a little squint in your eye.

Ready to join up?

Acknowledgments

There are so many people who have formed our paths, to list all of you could fill a chapter. Here are a few who provided significant boosts—and please forgive us if we've accidentally left you out; all our brains are scrunched here at the moment.

In chronological order: Jerry Montgomery, Charles Dygert, Charles Avery, John Morin, David Egloff, David Benzing, Richard Levin, David Merrell, Robert G. McKinnell, Donald Gilbertson, Philip Regal, Charles Burnham, Fritz Midelfort, Larry Geno, Mary Lewis, Martin and Alzora Rutter, Paul Pritchard, Brad Stanback, Don Willeke, Hank Roberts, Robin Ramsey, Wang Hong Wen, Professor Zhang (Hubei Academy of Agriculture), David Bakken, Donald Lawrence, John Herrington, Jan Joannides, the Minnesota Department of Agriculture, Jerry Henkin, Mark Decot, Matthew Nowak, Erv Oelke, William O. Hunt Jr., Megan Rutter, and Perry and Gina Rutter.

In addition, our special thanks to our publisher Chelsea Green for believing in us and all the teams' hard work; to senior editor Makenna Goodman for tracking us down and convincing Chelsea Green, and to our editor Fern Marshall Bradley for her terrific job in translating our not infrequently clueless scientific jargon into English.

—PHILIP RUTTER

To my parents, Paul and Gwen Wiegrefe, who nurtured the plant geek in me from an early age; Rudy Knutson, who first encouraged my scientific aspirations; Drs. Ray Guries and Ed Hasselkus, who patiently guided me through graduate school; Dr. George Ware, a great naturalist and mentor; and Jim Humbert, whose encouragement and generosity made my own nut farm a possibility.

—SUSAN WIEGREFE

Besides my co-authors and those they have thanked, I would like to particularly acknowledge the support provided in various ways over the years by Mary Lewis and Carl Alstad, Perry and Gina Rutter, Ruth Lewis, Suzanne and John Peters, Dr. Roger Quinn, Dr. Roy Ritzmann, the CWRU Biorobotics Lab, Sandra Albro, Swarthmore College, Sixteen Feet, Sara West, Megan Rutter, the stockholders in Badgersett Research Corporation, Ken Heidlebaugh, the harvest and planting volunteers helping on the farm, and pretty much everybody else, but especially Sabina Suzanne Peters-Daywater.

—BRANDON RUTTER-DAYWATER

Notes

Introduction

1. United Nations Conference on Trade and Development, *Wake Up before It Is Too Late: Trade and Environment Review 2013*, http://unctad.org/en/PublicationsLibrary/ditcted2012d3_en.pdf.

Chapter 1. Hazels, Hybrid Hazels, and Neohybrid Hazels

1. David B. Lobell, Wolfram Schlenker, and Justin Costa-Roberts, "Climate Trends and Global Crop Production since 1980," *Science* 333, no. 6042 (July 2011): 616–20.
2. United Nations Food and Agriculture Organization, "Global Agriculture towards 2050," High Level Expert Forum: How to Feed the World in 2050, October 12–13, 2009, Rome, Italy, fao.org/fileadmin/templates/wsfs/docs/Issues_papers/HLEF2050_Global_Agriculture.pdf.
3. John Doebley, "The Genetics of Maize Evolution," *Annual Review of Genetics* 38 (2004): 37–59.
4. H. M. Pellett, D. D. Davis, J. L. Joannides, and J. J. Luby, *Positioning Hazels for Large-Scale Adoption*, a report prepared for the Minnesota Agricultural Research Institute by the University of Minnesota's Center for Integrated Natural Resources and Agricultural Management, 1998.

Chapter 2. The Mortal Sins of Modern Agriculture

1. P. A. Rutter, "Reducing Earth's 'Greenhouse' CO_2 through Shifting Staples Production to Woody Plants," *Proceedings of the Second North American Conference on Preparing for Climate Change* (Washington, DC: Climate Institute, 1989): 208–13.
2. J. R. Smith, *Tree Crops: A Permanent Agriculture* (New York: Devin-Adair, 1953).
3. Ross Mars, *The Basics of Permaculture Design* (White River Junction, VT: Chelsea Green Publishing, 2005), 1.
4. "Backpack History—How and When Did They Come into Being?" Fabric and Handle, September 13, 2011, http://www.fabric-and-handle.com/articles/backpack-history---how-and-when-did-they-come-into-being.
5. L. Bouke van der Meer, *Ostia Speaks: Inscriptions, Buildings and Spaces in Rome's Main Port* (Leuven, Belgium: Peeters, 2012).

6. Craig Cox, Andrew Hug, and Nils Bruzelius, "Losing Ground," Environmental Working Group, April 2011, http://static.ewg.org /reports/2010/losingground/pdf/losingground_report.pdf.

7. N. S. Robins, ed., *Groundwater Pollution, Aquifer Recharge and Vulnerability* (London: Geological Society Special Publications 130, 1998).

8. Bryan Walsh, "This Year's Gulf of Mexico Dead Zone Could Be the Biggest on Record," *Time* (June 19, 2003), http://science.time .com/2013/06/19/this-years-gulf-of-mexico-dead-zone-could-be -the-biggest-on-record.

9. "Water Facts and Figures," International Fund for Agricultural Development (IFAD), n.d., http://www.ifad.org/english/water /key.htm.

10. "All Dried Up," *Economist* (October 12, 2013), http://www .economist.com/news/china/21587813-northern-china-running -out-water-governments-remedies-are-potentially-disastrous-all.

11. Alan D. Blaylock, *Soil Salinity and Salt Tolerance of Horticultural and Landscape Plants,* University of Wyoming Cooperative Extension Service, October 1994.

12. "IPCC Fourth Assessment Report: Climate Change 2007," Intergovernmental Panel on Climate Change, https://www.ipcc.ch /publications_and_data/ar4/wg1/en/spmsspm-human-and.html.

13. G. T. Miller, *Living in the Environment*, 12th ed. (Belmont, CA: Wadsworth/Thomson Learning, 2002).

14. Rutter, "Reducing Earth's 'Greenhouse' CO_2."

15. Linda Serck, "Pangbourne Sprays to Kill Oak Processionary Caterpillar 'Dangerous,'" *BBC News Berkshire*, June 5, 2013, http://www.bbc.com/news/uk-england-berkshire-22755526.

16. Lars Gamfeldt, Helmut Hillebrand, and Per R. Jonsson, "Multiple Functions Increase the Importance of Biodiversity for Overall Ecosystem Functioning," *Ecology* 89, no. 5 (May 2008): 1223–31.

17. E. O. Wilson, *The Creation: An Appeal to Save Life on Earth* (New York: W. W. Norton, 2006).

18. Oregon State University Extension Service, *Control of Moss and Lichens on Fruit Trees,* August 2000.

Chapter 3. The World Hazelnut Industry

1. "Marketing Order 982: Oregon and Washington Hazelnuts," USDA Agricultural Marketing Service, March 6, 2012, http://www.ams.usda.gov/AMSv1.0/ams.fetchTemplateData.do ?template=TemplateN&page=FVMarketingOrderIndex Hazelnuts.

2. Nancy K. Osterbauer, "Genetic Variability in the Eastern Filbert Blight Pathosystem," (PhD dissertation, Oregon State University, May 9, 1996), http://ir.library.oregonstate.edu/xmlui/bitstream /handle/1957/34613/OsterbauerNancyK1996.pdf?sequence=1.

3. Ramón Estruch et al., "Primary Prevention of Cardiovascular Disease with a Mediterranean Diet," *New England Journal of Medicine* 368 (April 2013): 1279–90; "Eat All the Nuts You Want" (blog entry), Badgersett Research Corporation, March 1, 2013, http://badgersettresearch.blogspot.com/2013/03/eat-all-nuts-you -want-really-really.html.

4. Nathan Gray et al., "FAO: Ukraine–Russia Tensions and Poor Weather Lead to Sharp Rise in Food Prices," Food Navigator, April 3, 2014, http://www.foodnavigator.com/Market-Trends/FAO -Ukraine-Russia-tensions-and-poor-weather-lead-to-sharp-rise -in-food-prices.

Chapter 5. Planting and the Establishment Period

1. P. A. Rutter, "Badgersett Research Farm: Projects, Goals, and Plantings." In *78th Annual Report of the Northern Nut Growers Association* (Lexington: University of Kentucky, 1987), 173–86.

Chapter 6. Pest Management

1. Michael Phillips, *The Apple Grower*, 2nd ed. (White River Junction, VT: Chelsea Green, 2005).

2. P.A Rutter and B. L. Rutter, "Eastern Filbert Blight: A Problem Solved," presented to 101st Annual Meeting of the Northern Nut Growers Association, print in preparation, 2010.

3. "Eastern Filbert Blight" factsheet, Cornell University Plant Disease Diagnostic Clinic, 2013, http://plantclinic.cornell.edu/factsheets /efilbertblight.pdf.

4. S. A. Katovich, A. S. Munson, J. Ball, and D. McCullough, "Bronze Birch Borer: Forest Insect & Disease Leaflet 111," http://www.na.fs .fed.us/spfo/pubs/fidls/bbb/bbb.htm.

5. Robert P. Wawrzynski, Vera A. Krischik, and Steven A. Katovich, *The Bronze Birch Borer and Its Management* (St. Paul: Minnesota Extension Service, University of Minnesota, 2009).

6. Philip Rutter, unpublished data, 2008.

7. L. W. Treadwell and R. H. Storch, "Observations on Phenology, Development, and Mortality of Larvae of the Hazelnut Weevil (*Curculis obtusus* [Blanchard]: Curculionidae) in Nuts of Beaked Hazelnut (*Corylus cornuta* Marshall: Betulaceae) in Thickets in Maine," *Journal of the New York Entomological Society* 105, no. 3/4 (1997): 221–29.

8. Leonard Gianessi and Ashley Williams, "Turkey #1 Hazelnut Producer Thanks to Insecticides," Crop Protection Research Institute, May 2012, accessed April 10, 2014, http://www.croplife.org/view_document.aspx?docId=3728.

9. "Notes on Nut Insects: Big Bud Mite of Hazelnut," Ontario Ministry of Agriculture and Food, http://www.omafra.gov.on.ca/english/crops/pub360/notes/nutbudmite.htm.

10. Vera Krischik and Doree Maser, "Japanese Beetle Management in Minnesota," University of Minnesota Extension, 2011, http://www.extension.umn.edu/distribution/horticulture/dg7664.html.

11. Ibid.

12. Edward C. Cleary and Scott R. Craven, "13-Lined Ground Squirrel," Internet Center for Wildlife Damage Management, 2005, icwdm.org/handbook/rodents/13linedgroundsquirrel.asp.

13. Ibid.

14. "White Footed Mouse," Penn State New Kensington, 2004, http://www.psu.edu/dept/nkbiology/naturetrail/speciespages/whitefootedmouse.htm.

Chapter 7. Managing Mature Hazels

1. Fred Madison, Keith Kelling, Leonard Massie, and Laura Ward Good, "Guidelines for Applying Manure to Cropland and Pasture in Wisconsin," University of Wisconsin–Extension Bulletin R-8-95-2M-E, 1995, http://www.fao.org/prods/gap/database/gap/files/373_MANU_GUI.PDF.

2. Carl J. Rosen and Peter M. Bierman, "Using Manure and Compost as Nutrient Sources for Vegetable Crops," University of Minnesota Extension, 2013, http://www.extension.umn.edu/garden/fruit-vegetable/using-manure-and-compost/.

3. Phillips, *Apple Grower*.

4. Julia W. Gaskin, "Best Management Practices for Wood Ash as Agricultural Soil Amendment," 2002, extension.uga.edu/publications/detail.cfm?number=B1142.

5. Judy Scott, "Wood Ashes Can Benefit Gardens and Lawns," last update December 9, 2010, extension.oregonstate.edu/gardening/node/1022.

6. "Why Use Poultry Manure?" Royal Horticultural Society, 2014, http://www.rhs.org.uk/advice/profile?pid=297.

7. Charles C. Mann, *1491: New Revelations of the Americas before Columbus* (New York: Knopf, 2005).

Chapter 9. Processing: From Harvest to Market

1. Wikipedia, "Henry Quackenbush," http://en.wikipedia.org/wiki/Henry_Quackenbush.

Chapter 11. Co-Products and Their Value

1. Coryloideae consists of *Corylus* (hazel), *Carpinus* (musclewood), *Ostrya* (hophornbeam), and *Ostryopsis*. K. O. Yoo and J. Wen, "Phylogeny of *Carpinus* and Subfamily Coryloideae (Betulaceae) Based on Chloroplast and Nuclear Ribosomal Sequence Data," *Plant Systematics and Evolution* 267, nos. 1–4 (2007): 25–35.
2. "Project 12057PJ Analytical Laboratory Report," Agricultural Utilization Research Institute, 2012.
3. Johannes Lehmann and Stephen Joseph, eds., *Biochar for Environmental Management: Science and Technology* (London: Earthscan, 2009).
4. James Bruges, *The Biochar Debate: Charcoal's Potential to Reverse Climate Change and Build Soil Fertility* (White River Junction, VT: Chelsea Green, 2009).
5. Composition Materials Company, "Walnut Shell Media," http://www.compomat.com/walnut-shell-blasting-media.
6. How Stuff Works, "8 Reuses for Nutshells," http://recipes.howstuffworks.com/reuse-nutshells1.htm.
7. Leslie Cole, "Peerless Pork," *Oregon Live*, June 23, 2009, http://www.oregonlive.com/foodday/index.ssf/2009/06/peerless_pork.html.
8. "New Source for Biodiesel Fuel," KAAL ABC 6 News coverage of Badgersett Research Corporation, 2007. Linked on http://www.badgersett.com/news/. Expired KAAL news story link was http://kaaltv.com/article/stories/S171140.shtml?cat=10219; link may be reactivated.
9. Y. X. Xu, M. A. Hanna, and S. J. Josiah, "Hybrid Hazelnut Oil Characteristics and Its Potential Oleochemical Application," *Journal of Industrial Crops and Products* 26 (2007): 69–76.
10. E. Hammond, "Identifying Superior Hybrid Hazelnut Plants in Southeast Nebraska," (Master's thesis, University of Nebraska-Lincoln, 2006).

Chapter 12. Neohybrid Hazels—Beyond Mendel

1. Wikipedia, "Hybrid Swarm," http://en.wikipedia.org/wiki/Hybrid_swarm.
2. George Ordish, *The Great Wine Blight*, 2nd ed. (London: Pan Macmillan, 1987).
3. Vadim Mozhayskiy and Ilias Tagkopoulos, "Guided Evolution of In Silico Microbial Populations in Complex Environments Accelerates Evolutionary Rates through a Step-Wise Adaptation," *BMC Bioinformatics* 13, supplement 10 (2012): S10.

Chapter 13. Climate Change, Resilience, and Neohybrid Hazels

1. Louis F. Pitelka, "Plant Migration and Climate Change," *American Scientist* 85, no. 5 (1997): 464.

2. Joe McAuliffe, "Ancient Creosote Bush Clones: A Trail of Multidisciplinary Discoveries," *Sonoran Quarterly* (2006), http://tinyurl.com/nso4999.

3. Paul V. A. Fine, Tracy M. Misiewicz, Andreas S. Chavez, and Rob Q. Cuthrell, "Population Genetic Structure of California Hazelnut, an Important Food Source for People in Quiroste Valley in the Late Holocene," *California Archaeology* 5, no. 2 (December 2013): 353–70.

4. "Sunrise, Sunset, Daylight in a Graph," ptaff.ca, 2005, http://tinyurl .com/m5229hh. This is the best, simplest set of graphs we could find. Most are much more complex and confusing; even these will take some time to understand.

5. P. E. Heilman and R. F. Stettler, "Genetic Variation and Productivity of *Populus trichocarpa* T. and G. and Its Hybrids: II. Biomass Production in a 4-Year Plantation," *Canadian Journal of Forest Research* 15, no. 2 (1985): 384–88; J. D. Ovington, D. Heitkamp, and D. B. Lawrence, "Plant Biomass and Productivity of Prairie, Savanna, Oakwood and Maize Field Ecosystems in Central Minnesota," *Ecology* 44, no. 1 (1963): 52–63.

6. Lehmann and Joseph, eds., *Biochar for Environmental Management.*

7. J. D. Ovington and D. B. Lawrence, "Comparative Chlorophyll and Energy Studies of Prairie, Savannah, Oakwood and Maize Ecosystems," *Ecology* 48, no. 4 (1967): 515–24.

8. C. P. Osborne and D. J. Beerling, "Nature's Green Revolution: The Remarkable Evolutionary Rise of C_4 Plants," *Philosophical Transactions of the Royal Society B: Biological Sciences* 361, no. 1465 (2006): 173–94.

9. Susanne von Caemmerer, W. Paul Quick, and Robert T. Furbank, "The Development of C_4 Rice: Current Progress and Future Challenges," *Science* 336, no. 6089 (June 2012): 1671–72, doi:10.1126/science.1220177.

10. G. Edwards and D. Walker, C_3, C_4: *Mechanisms, and Cellular and Environmental Regulation, of Photosynthesis* (Berkeley: University of California Press, 1983).

11. Anders Fischer, "Coastal Fishing in Stone Age Denmark— Evidence from Below and Above the Present Sea Level and from Human Bones." In N. Milner, O. E. Craig, and G. N. Bailey, eds., *Shell Middens in Atlantic Europe* (Oxford, UK: Oxbow Books, 2007), 54–69, https://www.york.ac.uk/archaeology/middens/ Chap%2005.pdf.

12. Oliver Rackham, *Woodlands* (London: Collins, 2010).

13. Kat Anderson, *Tending the Wild: Native American Knowledge and the Management of California's Natural Resources* (Berkeley: University of California Press, 2006).

14. Fine, Misiewicz, Chavez, and Cuthrell, "Population Genetic Structure of California Hazelnut."

15. Tom Wahl, personal communication, 2001.

16. Jeff Olsen, "Nut Growers Handbook," Oregon Hazelnuts, n.d., http://www.oregonhazelnuts.org/growers-corner/grower -handbook.

17. Gregory Miller, "Pre- and Post-Harvest Changes in Chinese Chestnuts: Implications for Mechanical Harvest and Storage." In *78th Annual Report of the Northern Nut Growers Association* (Lexington: University of Kentucky, 1987), 81–87.

18. H. P. Collins et al., "Soil Carbon Pools and Fluxes in Long-Term Corn Belt Agroecosystems," *Soil Biology and Biochemistry* 32 (2000): 157–68. This article is technically dense, but good work.

19. Lehmann and Joseph, eds., *Biochar for Environmental Management.*

Chapter 14. The State of the Crop

1. Pellett, Davis, Joannides, and Luby, *Positioning Hazels for Large-Scale Adoption.*

2. American Soybean Association, "90 Years of Soybean Progress with ASA," 2010, http://soygrowers.com/wp-content/uploads /2013/04/ASA-90th-Insert-w.pdf.

3. N. J. C. Zerega, D. Ragone, and T. J. Motley, "The Complex Origins of Breadfruit (*Artocarpus altilis*, Moraceae): Implications for Human Migrations in Oceania," *American Journal of Botany* 91, no. 5 (2004): 760–66.

Resources

Further Reading

We have cited a great many excellent books, papers, and websites in the notes, and we don't generally repeat them here. This is rather a list of books that we find excellent or potentially life changing. Typically they have enormous bibliographies of their own, which you can also winnow. Be aware that their actual relevance likely extends far beyond the category we have put them in.

UNDERSTANDING HUMANS
Ernest Thompson Seton, *Two Little Savages* **(Doubleday, 1903);**
540 pages

> Though he initially presented this story as fiction for young boys, Seton eventually admitted it was entirely autobiographical. It's the true story of the summer that led to the creation of the Boy Scouts. Seton wrote the first *US Boy Scout Handbook*, and was the first US chief scout. When I asked a friend, now a prominent ornithologist, "Have you ever read this?" he responded with an explosive "It changed my entire life!" It's also full of entirely practical information and diagrams.

Philip J. Regal, *The Anatomy of Judgment* **(University of Minnesota**
Press, 1990); 368 pages

> This book—described as "a humanist cult classic"—discusses the human mind from the perspectives of neural physiology and anatomy, anthropology, history, science, religion, and more.

Michael Pollan, *The Omnivore's Dilemma* **(Penguin Press, 2006);**
450 pages

> Food and food systems from the human perspective. Eye opening for everyone, with a good discussion of Joel Salatin's farm.

Joel Salatin, *Everything I Want to Do Is Illegal* **(Polyface Press, 2007);**
338 pages

> Read his other books, too. Salatin is rock-solid, top to bottom.

UNDERSTANDING ECOLOGY, GENETICS, AND EVOLUTION
David J. Merrell, *Ecological Genetics* **(Longman Group, 1981);**
500 pages

> Graduate-level textbook, one of the first in the field. Even just scanning it will open your eyes.

Michael L. Rosenzweig, *Species Diversity in Space and Time* (Cambridge University Press, 1995); 436 pages

> Rosenzweig is one of the leaders in the academic study of biodiversity, how it works, and why it's critically important.

Michael Pollan, *The Botany of Desire* (Random House, 2001); 273 pages

> Pollan's first book is a look at plant breeding from a non-biologist's perspective. We need plants *and* people.

DEALING WITH CLIMATE CHANGE

Eliot Wigginton, editor, *The Foxfire Book* (Anchor Books, 1972); 384 pages

> A collection of conversations with backwoods grandparents, this would appear to be about old-time ways—but it's actually a tremendous resource on both simple life and community. This is only the first of a large series of books. The Foxfire program is still active, and *Foxfire Magazine* is still in production: foxfire.org.

Michael L. Rosenzweig, *Win–Win Ecology* (Oxford University Press, 2003); 211 pages

> Written for the general audience, this is an outline of practical action to preserve biodiversity and ecological health in times of human population growth. It achieved something no other book I know of has: It got top book reviews in both *The New York Times* and *Science*.

Sharon Astyk, *Depletion and Abundance: Life on the New Home Front* (New Society Publishers, 2008); 288 pages

> An intelligent, educated, authoritative, and lucid discussion of our possible future.

Rob Hopkins, *The Transition Handbook* (Chelsea Green Publishing, 2008); 239 pages

> The most successful and intelligent movement worldwide at building new communities centering on durable resilience and decreased reliance on fossil fuels. If you're thinking of launching a community, study the Transition Towns Movement first. Abundant web presence and other publications available.

Risa Bear, *Starvation Ridge* (LULU Press, 2014); 462 pages

> This is one of the most realistic, true to life portrayals of a possible world after global collapse I know of. Eye opening, deeply detailed, and very good reading.

Carl N. McDaniel, *At the Mercy of Nature* (Sigel Press, 2014); 180 pages

> A unique analysis of the survival of the Shackleton endurance expedition by a prominent ecologist, with extensive reference to what our species must and must not do if we are to survive climate change. Endorsed by E. O. Wilson, Bill McKibben, and Wes Jackson.

UNDERSTANDING TREES AND WOODY PLANTS

Oliver Rackham, *Woodlands* **(Harper Collins, 2006); 508 pages**

> Stunningly in-depth work on the history of humans and the uses of wood in Europe. Would win the Nobel Prize in Forestry—if there were one—and readable, too.

Shawn A. Mehlenbacher, "Hazelnuts," chapter 17 in *Genetic Resources of Temperate Fruit and Nut Crops, Volume 2,* **edited by J. N. Moore and J. R. Ballington Jr. (International Society for Horticultural Science, 1991); pages 791–838**

> A good start on the entire biology of hazels, this chapter includes full chemical and nutritional analyses of three different cultivars for an appreciation of typical range of variation. Neohybrids were not described when this was written, and are not included. (You might want to note who wrote chapter 16, "Chestnuts" . . .)

Growing Information

STATE UNIVERSITY EXTENSION EDUCATORS AND COUNTY AGENTS

Both states and national agencies provide agricultural consultants to farmers, and are rarely hard to find. If your phone book doesn't help, ask your farmer neighbors. These folks should be your first go-to resource to learn the details of growing in your area. From soil types to fertility needs and where to get your soil tested, these folks know your local area in detail. Often they will come and walk your farm with you, answering your questions with specific local information. They are not likely to be familiar with hybrid hazels, but they are also usually interested, and will be helpful.

OLD FARMERS

It will require an investment of your time to find good local old farmers and convince them to share what they know. Yet there is no more rewarding investment you could make.

MASTER GARDENERS

Many places have a certified Master Gardener program. These folks can be wonderful and helpful. Beware the occasional "know-it-all" type—who doesn't know it all, and doesn't know it.

BADGERSETT ANNUAL SHORT COURSE

badgersett.com/infoproducts/shortcourse

Every year since 2006 we have taught an intensive two- to three-day short course near our farm in southeastern Minnesota. Everything in this book is enlarged on and brought up to date during the course, and there is abundant time to ask questions, see plantings, and talk with other growers, old and new. Quite a few people re-attend periodically to get updated. Visit our short course webpage for information on dates, details, and so forth. The course now varies from year to year, with a pre-announced focus.

We are attempting to make this course available in other regions as soon as we can. Send us an email (info@badgersett.com) if you are interested in organizing one in your area. Usually we need at least 20 guaranteed paying students to consider the travel needed.

BADGERSETT SHORT COURSE DVDS

Purchase information on this page: badgersett.com/catalog

We've only recently been able to offer DVDs of the 2011 and 2014 Badgersett Short Courses, recorded carefully at the events and edited by Dr. Brandon, who has also produced videos for DARPA. These represent many hours of seminars with hundreds of images and animations. Each year the courses are somewhat different, and to improve communications we have begun rotating which of us present the topics from year to year—our perspectives are always slightly different. The one-hour introduction from 2011 is on YouTube as a preview: youtube.com /watch?v=KpJR2yfLUU0.

Nut Growers Associations

nutgrowing.org

The first thing I did when I decided to seriously tackle nut crops was to join the Northern Nut Growers Association (NNGA), now over 100 years old. It was the right decision, and eventually I served as president.

Many states have their own nut growers associations independent of the NNGA; many growers belong to both. They all offer a great trove of information and experience, both extremely valuable. The membership is about 80 percent "amateur," in the best and original sense of that word, and 20 percent academics. There is usually some friendly (and educational) friction between the two groups. Sometimes the expertise only holds up locally.

Find your state group and join; if you don't have a state group, start one. Go to the meetings, visit the plantings, listen to everything—and take everything with a grain of salt, amateur or professional.

Tools and Supplies

Yes, you can buy gardening tools and supplies at Walmart—but if you're aiming at a business and a farm, you'll be better served by finding professional sources. Stores and national chains run the gamut in their customer target from (#1) Suburban Farmer Wannabes to (#2) Hobby Farmers to (#3) Real Farmers. Prices for #1 will be double those for #3, with a tool life span of one to two years. Prices for #2 will be triple those for #3, with a tool life span of four to five years. Prices for #3 will be the lowest you can find, and tool life span will be from 10 to 20 years. Find #3; you may need to ask a real farmer where it is. The #3 stores are an entire education in themselves, and usually a link to community also.

PROFESSIONAL LANDSCAPE AND FORESTRY TOOLS
Landscaping and forestry are very large industries, with many highly specialized tools you are unlikely to find anywhere else. These are major professional suppliers, easily available on the Internet. These suppliers have everything. Beware of their catalogs; you can get sucked in for hours.

Forestry Suppliers
forestry-suppliers.com; 800-647-5368; 205 West Rankin Street, PO Box 8397, Jackson, MS 39284-8397
Ben Meadows
benmeadows.com; 800-241-6401; PO Box 5277, Janesville, WI 53547-5277
A. M. Leonard
amleo.com; 800-543-8955; 241 Fox Drive, Piqua, OH 45356

SMALLHOLDER EQUIPMENT
Earth Tools
earthtoolsbcs.com; 502-484-3988; 1525 Kays Branch Road, Owenton, KY 40359
> We own two of their diesel walking tractors with multiple implements, with many years of hard work already on them. No sign of wearing out. We will buy our next one here. First-rate service and expertise, Real Farmer quality.

Timeless Tools
timelesstools.co.uk
> Professional-grade wood management tools, many hand-forged by top craftsmen trained inside a toolmaking tradition that is many hundreds of years old—probably thousands. Someone here on the farm overheard me muttering, "Boy, I would sure like to have a Yorkshire billhook to do this job with . . . "—so I got a Yorkshire billhook for Christmas, from Timeless Tools. I have a US patent in metallurgy (in my spare time), and I am in awe of the master craftsmanship in every aspect of this tool. No, you can't find these in the United States.

Plant Nurseries for Hybrid Hazels with Good Breeding Practices
As you know, Badgersett launched all the current efforts to develop new hybrid hazel crops for the North American Upper Midwest and beyond—including growers who prefer not to mention that. Please know that we are 100 percent aware of these folks, and know how they currently get the seed for what they sell. If they are not listed below, it's because we

believe their specific processes and seed will not and cannot lead to any crop progress, and that you will lose your money and time. We are aware of several people working to produce reliable neohybrid nursery stock, something we strongly believe is a good idea. More good plant sources are in the pipeline.

Badgersett Research Farm

badgersett.com/plants/#hazels; 888-557-4211; 18606 Deer Road, Canton, MN 55922

> We have been testing and keeping extensive long-term performance records on hazels since 1978, and breeding neohybrid hazels using both fully controlled crosses and semi-controlled crosses since 1988. The hazels you buy from us are guaranteed resistant to EFB, and have been grown from 100 miles north of Edmonton, Alberta, to Georgia. Our seed is selected with utmost care, much of it handpicked by our science staff, and with individual parent identity preserved. We are also the only nursery currently selling seedlings from machine-picked hazels, also carefully selected. We will help you succeed with your planting to our best ability; just call or email info@badgersett.com.

Red Fern Farm

info@redfernfarm.com; 319-729-5905; 13882 I Avenue, Wapello, IA 52653

> Tom and Kathy incorporated some Badgersett neohybrid hazel genetics into their own stock two decades ago, with good results. They have also been very active in organizing and supporting local nut growers. Excellent people.

Oikos Tree Crops

oikostreecrops.com; 269-624-6233; PO Box 19425, Kalamazoo Township, MI 49019

> Not a specialist in hybrid hazels, but providing good stock and expertise for many different tree crops. Their hazels are *not* descended from ours, but they are carefully improved for EFB resistance and cold tolerance, using long-term testing; they are worth a try. Good honest business folks.

Twin Ponds

> This is the business of the major Illinois neohybrid planting. For the first time in 2015, they will be producing about 8,000 neohybrid tubelings from advanced neohybrids; seed parents selected by Badgersett Research Corporation (BRC) research staff. Current plans for this first year call for orders and shipping to be handled through Badgersett; email info@badgersett.com. In cooperation with BRC.

Z's Nutty Ridge LLC

Cortland County, New York

> Not yet producing stock for sale, but coming soon—possibly in 2015. Unlike other nurseries, this establishment has a plan to produce

tissue-cultured clonal hazel plants selected in New York State from their 20-year-old planting of BRC neohybrids. This can be a major step forward for neohybrid hazel growers. While we don't expect to recommend planting all-cloned hazels anytime soon, having a significant proportion of clonal hazels can be a major advantage in multiple ways. Among them are more uniform production and quality; in addition, the clones provide a critical baseline to compare new seedlings with. In cooperation with BRC.

Index

Note: Page numbers in *italics* refer to photographs and figures; page numbers followed by *t* refer to tables.

abrasives, from hazel shells, 191
abscission, 153–54, *153*
acclimation of tubelings, 71, 88
acres in place, 78, *79t*
acres to harvest, 78, *79t*
acres to plant, 78, *79t*
acres to pre-plant, 78, *79t*
activated charcoal, from hazel shells, 191
agroforestry, 24. *See also* woody agriculture
air pollution, 30–32
alfalfa cover crops, 98
allelopathy, 14
allergy concerns, 180
alternate bearing effect, 40
American Chestnut Foundation, 4
American hazels
  characteristics of, 6, 9, *9*, 13, 16, *16*
  flavor of, 170
  genetic crossing with other types, 200–201
  mutualistic relationship with EFB, 107, *108*
  poor latitude adaptation, 215–16
  range of, 12–13, *13*
  role in neohybrids, 12–14
American Soybean Association, 229
A. M. Leonard, 247
animal feed co-products, 193, 194
animal manure, as fertilizer, 139–140

animal pest management. *See also* browsing management; *specific types of animals*
  direct seeding, 63–64
  mature plantings, 144
  tubelings, 71–72
Anisogramma anomala. See Eastern filbert blight (EFB)
anthracnose, 110–11, *110*
aphids, 122
augers for planting holes, 69
Australia, hazelnut production, 44
Azerbaijan, hazelnut production, 44

backcrossings, 205
backpack belts, 25
Badgersett Research Farm
  breeding cycle, 206–8
  contacting, 102, 248
  current status, 17–21, 53, 225–27, *226*
  history of, 3–6
  short course in woody agriculture, 203, 245–46
  visiting, *18*, 227
bare-root dormant hazels
  crowns, 62–63, *63*
  planting checklist, *89t*
  receiving and holding, 88
  tubelings, 62, 70
bare-root stock, tubelings vs., 60
barriers to genetic crossing, 200–201
barriers to risks, 64, 65, *65*, 72
bathtub drains, 30, 220

beaked hazels
  California hazel subspecies, 215
  genetic crossing with other types, 200–201
  husk of, 14, 16, *16*
  mutualistic relationship with EFB, 107, *108*
  poor latitude adaptation, 215–16
  range of, 14, *14*, 15
  role in neohybrids, 14–15
bear control, 133, 162
beating bushes, 157, 158, 159
BEI blueberry-picking machines, 157–58, *158*
beneficial organisms, 106, *106*, 122, 144
Ben Meadows, 247
Bergen, John, *96*
best by dates, 170
Bhutan, hazelnut production, 44
big bud mites, 116–18, *116*, *117*, 207
biochar, 190–91, 216, 222
biodiesel production, 194–95, *195*, 220, 223
biodiversity. *See also* genetic diversity
  crop diversity, 51
  ecosystem complexity benefits, 32–33, 35, 105–6
  habitat/biodiversity islands, 64, *65*
  modern agriculture role in loss of, 32–35
  restoration of, 33–35

trade-offs with cloning advantages, 147–48
biological carbon cycle, 222
biomass production
as appropriate measure of crop systems, 31–32
chlorophyll measurement vs., 216–17, *217*
uses of, 53, 188, 190
bird control, 130–32, *130*, 150, 155
bitterness of nuts, avoiding, 170
black bears, 133
blading, in ground preparation, 69–70
blasting (big bud mites), 116, *116*
blight. *See* Eastern filbert blight (EFB)
blocks of hazels, 51
blueberries, as model for hazelnuts, 76
blueberry-picking machines, for harvesting, 157–58, *158*
blue jays, 131, 155
boiling water model of climate change, 212–13
book resources, 243–45
Bosnia, hazelnut production, 44
bottleneck plants in selection process, 204–5
boundaries, field, 64, 128, 147
breadfruit cultivation, 231
breeding considerations. *See* genetics
brown-ripe stage, 154, *154*
browsing management
as cost category, 79, 80–81*t*, 83, 83*t*
egg spray, 99, 101, 132, 133, 155
human presence as deterrent, 103
immediately after planting, 92–93
for mature plantings, 144
year 1 requirements, 99

year 2 requirements, 100–101
building materials, hazelwood for, *178*, 188–89, *189*, 218–19
bulb planters, 89, 90, 95–96, *95, 96*
bull genetic charts, 20
burdock control, 149
Burnham, Charles, 4
bush beaters, 157, 158, 159
bush growth form
American hazels, 13
beaked hazels, 14
European hazels, 15
neohybrid hazelnuts, 15, 17
tree form vs., 6, 188
Turkey production, 40, *41*
business plans, 76–77

C4 vs. C3 plants, 32, 217–18
cake box huskers, 164–65, *164*
California, quarantine on imported hazels, 43–44
California hazels, 215
calluses, Eastern filbert blight, 109, *109*
Canada, hazelnut production, 185
cankers, Eastern filbert blight, *108*, 109
carbon cycle, biological vs. geological, 222
carbon dioxide
C4 vs. C3 plants, 32, 217–18
modern agriculture contribution, 31
whole crop carbon measurement, 216–17, *217*
woody agriculture mitigation potential, 32, 191, 221–22
care levels
defined, 77–78
expected crops vs., 77–78, 77*t*
first year requirements, 49
income projections vs., 83–85, *84*

of plant divisions, 147
survival rates and productivity vs., 87, 88
caterpillars, 120–21, *121*
catkins
American hazels, 13
beaked hazels, 15
big bud mite infestations, 117
function of, 8, *11*
as nutrient status indicator, 138
cats, for pest control, 126, 128, 133
cattle raising with hazelnut culture, 52
Cecropia caterpillars, *121*
charcoal, from shells, 190–91. *See also* biochar
chemicals, from husks, 193–94, *193*
chestnut hybrid swarm, 201
chicken raising with hazelnut culture. *See* poultry raising with hazelnut culture
Chile, hazelnut production, 44
China, hazelnut production, 44–45
chipmunk control, 127–28, *127*, 149, 155
chlorophyll measurement, 216–17, *217*
cleaning nuts, 83, 83*t*, 165–67, *166*
climate
moderation benefits of neohybrids, 221–22
site requirements and, 55–56
climate change, 211–223
adaptability of seed-grown vs. clonal hazels, 57
book resources, 244
current hazelnut supply vulnerabilities, 45
mitigation with biochar, 191
need for adaptation, 211–13
neohybrid advantages, 213–223

clonal vs. seed-grown hazels, 57–59, 58t, 147–48, 205
clover cover crops, 98
cold hardiness, selecting for, 205, 207, 208
cold hours to break dormancy, 62
color coding plants, 145
commercial hazelnut production. See world hazelnut industry
commercial-kitchen grade facilities, 175
commodity market development, 45, 181–83
communication, importance of, 231–32
community, importance of, 232
companion crops, 36, 51, 140
composted manure, as fertilizer, 139
composting, of hazel shells, 192
confounding variables, 68–69
Conservation Reserve Program (CRP) planting, 63, 72, 73
consumables costs, 82, 82–83t
control groups (experimentation), 68–69, 231
coppicing
  deliberate killing of plants, 145
  during drought, 102
  history of, 7–8, 218
  income from, 83, 178, 179, 187
  machine-aided, 53, 143–44
  for productivity and vigor, 66, 142–44, 142
  scheduling, 142–43, 187
  superior qualities of neohybrids, 7–8, 218–19
co-products, 187–195
  husks, 192–94, 193
  as neohybrid advantage, 53, 220–21
  shells, 190–92
  sub-food-grade nuts, 194–95, 195
  wood, 83, 187–190, 189

corn
  development from teosinte, 2, 19, 201–2, 202, 228
  fertilizer recommendations adapted to hazelnuts, 136
  genetic diversity vs. neohybrids, 197
  photosynthetic activity vs. neohybrids, 217–18
  transition to hybrid version, 18–21, 229
  unrealized goal of perennial form, 3
  whole crop carbon measurement, 216
Cornell University, 231
Coryloideae subfamily, 188
Corylus americana. See American hazels
Corylus avellana. See European hazels
Corylus colurna. See Turkish hazels
Corylus cornuta. See beaked hazels
cosmetic abrasives, from hazel shells, 191
cost calculations
  acreage and rows for, 78, 79t
  activity details, 81–83, 82–83t
  effects of care on expected harvest, 77–78, 77t
  establishment costs, 57, 75–85
  seed costs, 64
  summary for example planting plan, 78–81, 80–81t
cover crops, 98
cracking nuts, 151, 174–76, 174, 175
craft uses of hazelwood, 178, 188–89, 189, 218–19
Crompton, Derek, 194
crop diversity, as goal, 51
crop index, 8

crowd-sourcing approach to domestication, 208–9, 209
crown tissue, 8, 147
crows, deterring, 131, 150, 155
culling plants, 66, 67, 67, 144–45
cultivation, for weed control, 96–97. See also weed control
cumulative expenses cost category, 80, 80–81t
cumulative gross income, in example plan summary, 80–81t, 81
cumulative net income, in example plan summary, 80–81t, 81
curing hazelnuts, 158, 161

day length adaptations, 215–16
decapitation of tubelings, 71, 93
decision to plant, 47–50, 49t
deer control, 97, 99, 101–2, 132, 155
deer mice control. See mice control
density of planting. See plant spacing
depth, planting, 90–91, 93
diameter of full-grown plants, 67
dibble spikes, 89
direct marketing, 179
direct seeding, 63–64
diseases, 107–11
  anthracnose, 110–11, 110
  Eastern filbert blight, 43–44, 107–9, 108, 109, 113, 205, 207, 208, 227–28
  mature hazelnut concerns, 144
disking, in ground preparation, 69
diversity. See biodiversity; genetic diversity
dividing plants, 147
dogs, for pest control, 103, 125, 128, 133, 198–99

domestication
of corn, 2, 19, 201–2, *202*, 228
crowd-sourcing approach to, 208–9, *209*
hybrid swarm genetics for, 201–6, *202, 203*
of symbionts, 198–99
of trees, 25–26
dormancy, cold hour requirements, 62
drought
effect on ripening, 154
establishment during, *91*, 102
modern agriculture contribution to, 35–36
water requirements in, 79, 94, 102
drying hazelnuts, 163–64
dry-ripe stage, 154, *154*
Dutch white clover cover crops, 98

early-ripening hazels, 58
Earth Tools, 247
Eastern chipmunk control. *See* chipmunk control
Eastern filbert blight (EFB)
association with stem borers, 113
history of, 43–44
neohybrid resistance to, 5, 107, 108, 109, 205, 207, 208
overview, 107–9, *108, 109*
susceptibility to, 43–44, 107, 227–28
east-west vs. north-south row placement, 67
economic considerations. *See also* cost calculations; income
avoiding expensive mistakes, 21
example plan summary, 78–81, 80–81*t*
financial modeling, 76–77

pricing of hazelnuts, 45, 181
ecosystem pest management. *See also* pest management
benefits of biodiversity, 32–33, 35, 105–6
cost considerations, 77
habitat/biodiversity islands, 64, *65*
for mature plantings, 144
overview, 48–49, 105–7
wildlife habitat practices, 54, 58, 92–93
edible oil crops, *24, 26*
EFB. *See* Eastern filbert blight (EFB)
egg spray, 99, 101, 132, 133, 155
energy production from hazels, 222–23. *See also specific types*
England, coppicing history, 142, 189, 218
enterprise budget tools, 76–77
The Environmental Working Group, 26–27
equipment. *See* tools and equipment
erosion concerns
example planting plan, 72
hazel performance, 220
during planting, 92
establishment period, 93–103
cost calculations, 57, 75–85
defined, 87
during extreme conditions, 94, 103–4
pocket gopher threat, 99, 100–101, 123
year 1, 93–99, *95–99*
year 2, 100–101
years 3 and 4 requirements, 101–2
Europe
coppicing history, 142, 189, 218
hazelnut market, 41, 44, 183–85
European hazels
characteristics of, 6, 16, *16*

climate adaptations, 55
coppicing of, 7–8
flavor of, 170
genetic crossing with other types, 201
range of, 15
role in neohybrids, 15
susceptibility to EFB, 107
Turkey production, 40
example planting plan
field layout details, 72–75, *72, 74*
summary of costs, harvest and income projections, 78–81, 80–81*t*
excellent care. *See also* care levels
cost and income projections, 84–85, *84*
defined, 77
exit holes, insect, 112, *112*, 114, *114*
experimental genetics, in planting mix, 58–59
experimentation
conflicting expert advice and, 103
control groups, 68–69, 231
mulching, 74–75
planning for, 68–69
expert advice, conflicting, 102–3
extension agents, 245
eye protection, 151

fabric mulches, 97, *99*
faggots, hazelwood, 188
fall coloration, 137–38, *138*
famine concerns, 35
FAO (United Nations Food and Agriculture Organization), 1, 39
FAOSTAT (UN FAO database), 39
farmers
as geneticists, 197–99
importance of involvement, 76, 93, 245

role in developing industry,
228–230
farmers markets, selling at, 179,
180, 189, 191
farm-generated propagation,
146–48
Farris, Cecil, 5
federal marketing order, Oregon
and Washington, 42
feed grade urea, as fertilizer, 94
Ferrero Company, 41, 184–85
fertility
of plants, 96, 113, 136, *136*,
144
of soil, 56, 191
fertilizers
adapting corn
recommendations to
hazelnuts, 136
as cost category, 79, 80–81*t*,
83, 83*t*
experiments with, 29–30, 68
mature planting
requirements, 135–140
need for, 225–26
organic vs. synthetic, 52–53,
136–37, 139–140
role in pest management, 105
timing of, 137
water contamination
from, 29
year 1 requirements, 94–96,
95, *96*
year 2 requirements, 100
field boundaries, 64, 128, 147
field layout, 64–68, *65*, *66*, *67*,
88
filazels (filbert-hazelnut cross),
12
filbertone, 161, 170
financial considerations. *See*
economic considerations
fire, evolutionary adaptations to,
7, 62, 142, 218
first year care requirements, 49
Fischbach, Jason, 76
flags, wire stake, 90, 152

flail mowing, 149
flavor of nuts
consistency of, 181
high-end market potential,
183
judging, 168–172, *171*, *172*
selecting for, 207
flea beetles, 120, *120*
flickers, 132
flood concerns, 36–37, 102,
219–220
florist's tape, 152
food crisis, depth of, 1–2, 198
food-grade cracking equipment,
175
food safety concerns, 175–76,
180
forestry machinery, for
coppicing, 143–44
Forestry Suppliers, 247
fowl raising. *See* poultry raising
with hazelnut culture
fox squirrels, 128–29
France, hazelnut production, 44
fuel
biodiesel production,
194–95, *195*, 220, 223
biomass production, 31–32,
53, 188, 190, 216–17, *217*
as cost category, 82, 82–83*t*
furniture uses of hazelwood,
178, 188–89, *189*

galls, stem borer, 112, *112*
garden rakes, 90
garlic mustard control, 148
Gates Foundation, 217
geese raising with hazelnut
culture, 51
Gellatly, Jack, 5
generation vs. selection cycle,
207
genes
defined, 202
numbers involved for
essential traits, 206–8, 215

selecting for desired traits,
203–6, *203*
genetic charts, 20
genetic diversity
in neohybrids, 12, 20, 118,
153, 197, 201–6, *203*,
214–18, *217*
Pacific Northwest vs. eastern
EFB fungi, 43–44
power of hybrid swarms,
202–6, *203*
geneticists, farmers as, 197–99
genetic mix, 57–58
genetics, 197–209
Badgersett breeding cycle,
17–18, 206–8
big bud mite resistance
selection, 117–18
book resources, 243–44
clonal vs. seed-grown hazels,
57–59, 58*t*
crowd-sourcing approach to
domestication, 208–9, *209*
culling decisions, 145
desirable characteristics, 8–9
domestication principles,
201–6, *202*, *203*
EFB resistance selection, 5,
107, 108, 109, 205, 207, 208
farmers as geneticists,
197–99
flavor, 168–69
growth forms, 17–18, 188
hybrid swarms overview,
199–201
neohybrid vs. hybrid
approach, 12
nut weevil resistance
selection, 115–16
origin of neohybrids, 10–15
reliability, 169
in-situ propagation, 146–48
stem borer resistance
selection, 113
as tool, 231
tracking separate batches of
nuts, 174

two traits at time, selecting for, 204, 207
wood harvesting traits, 187, 188
geological carbon cycle, 222
Georgia (country), hazelnut production, 44
germination, 35, 146
gloves, 151
glyphosate herbicide, 97
goals for planting, 50–53, *51*, 54
good care. *See also* care levels
cost and income projections, 83–84, *84*, 85
defined, 77
gopher control. *See* pocket gopher control
Gordon, John, 5
grading nuts, 172–74
grass and legume companion crops, 36, 51, 140
grasshoppers, 121–22, *122*
gray squirrels, 128–29
grazing. *See also specific types of animals*
manure benefits, 139–140
for pest control, 133, 149–150, *150*
planning for, 52, 80
year 2 cautions, 100
Greece, hazelnut production, 44
greenhouse gases, 31, 32, 211
green-ripe stage, 153–54, *154*
gross income, in example plan summary, 80–81*t*, 81
ground marking paint, 90
ground preparation. *See also* soil
as cost category, 79, 80–81*t*, 82, 82–83*t*
overview, 69–70, *70*
subsoiling during, 56, 69
ground squirrels, 126–27
Growing Nuts in the North (Weschcke), 5
growth forms. *See* bush growth form; tree growth form

guinea fowl raising with hazelnut culture, 51

habitat/biodiversity islands, 64, *65*
hammermill-type huskers, 164–65, *164*
hand harvesting
as cost category, 83, 83*t*
details of picking nuts, 156–57
judging ripeness and, 154–55
plant spacing for, 67
of seed, 64
in Turkey, 39, *41*
hardiness zones, 55
hares, pest management, 133
harvesting, 151–162. *See also* hand harvesting; machine harvesting
checklist for, 151–52, *152*
costs of, 79, 80–81*t*
details of picking nuts, 156–160, *158*, *159*
expected crops over time, 77–78, 77*t*
nut theft, 151, 155
timing, 152–55, *153*, *154*
transport and storage of nuts, 161–62, *162*
in Turkey, 39, *41*
harvest percentage testing, 159
harvest transport bags, 151, *152*, 156
Hayes, Sean, *96*
hazelberts (hazel-filbert cross), 12
hazelnuts, overview, 6–9, *7*, *8*, *9*. *See also specific types*
health of plants. *See* fertility; plant vigor
heat concerns, 94, 102
height of plants, 6, 17–18
Hemenway, Toby, 24
herbicide concerns, 55, 65, *65*, 95, 97, 103

heterozygosity, 197
high-end markets, 183–85, *184*
hilum, 14–15, 152, 153, *153*
HMQ nutcrackers, 174–75, *174*
hoeing, 90, 92, 100
hope, importance of, 223
horse raising with hazelnut culture, 52, 140, 149–150
humor of nut farming, 59
hunger concerns, 1–2, 35, 36, 213
hurdles of hazelwood, *178*, 189, *189*, 218
husking machines, 163, 164–65, *164*, 227
husking nuts, 83, 83*t*, 163–65, *164*
husk of hazelnuts
American hazels, 13, 16, *16*
beaked hazels, 14, 16, *16*
breeding considerations, 9
co-products from, 192–94, *193*
European hazels, 16, *16*
neohybrid hazelnuts, 16, *16*
ripeness and color of, 154
hybrid corn, 18–19, 229
hybrid hazelnuts. *See also* neohybrid hazelnuts
barriers to genetic crossing, 200–201
neohybrids vs., 11–12, 199
premature optimism about, 227–28
Weschcke, Carl work, 5
hybrid poplar, whole crop carbon measurement, 216
hybrid speciation, 200–201
hybrid swarms, 10, 199–201
hydrated lime application, 141

ice traction, with hazel shells, 192
Illinois planting, 118, 226–27, 248
income
co-products, 187

example planting plan summary, 78–81, 80–81t
excellent care projections, 84–85, *84*
good care projections, 83–84, *84*
purse-nurse crops, 76, 80, 80–81t
wood harvesting, 83, *178*, 179, 187
industrial abrasives, from hazel shells, 191
infilling established plantings, 146
insect pests
big bud mites, 116–18, *116*, *117*
caterpillars, 120–21, *121*
flea beetles, 120, *120*
grasshoppers, 121–22, *122*
hazelnut aphids, 122
husking before transport concerns, 158
Japanese beetles, 118–19, *119*
in mature hazelnuts, 144
nut weevils, 114–16, *114*, *115*
potential husk-based control of, 193–94
stem borers, 111–13, *112*, *113*
in-situ propagation, 146–48
involucres. *See* husk of hazelnuts
Io moth caterpillars, *121*
Iowa State University, 26–27
Iran, hazelnut production, 44
irrigation. *See* watering
Italy, hazelnut production, 39, 41

Japanese beetles, 118–19, *119*

Kelty, Dick, 25
kernel defects, 171–72, *172*
killing plants, difficulty of, 145

kites, for predator simulation, 126, *126*, 131, 132, 150, 155

labor costs, 80, 82, 82–83t
lace leaf appearance, 120, *120*
landscape considerations, 54–55
landscape fabric mulches, 97, 99
landscaping uses of hazel shells, 192
land tenure considerations, 49
late-ripening hazels, 58
latitude adaptations, 215–16
Lawrence, Don, 216–17
leaching, of seed, 146
leaves
anthracnose damage, 110–11, *110*
of beaked hazels, 14
caterpillar damage, 121, *122*
flea beetle damage, 120, *120*
Japanese beetle damage, 118, *119*
removing during drought, 102
visual assessment of plant nutrient status, 96, 137–38, *137*, *138*
legume companion crops, 36, 140
lime application, 141
Littau blueberry-picking machines, 157
livestock in hazelnut culture. *See specific types of livestock*
local climate moderation benefits of hazels, 221

machine-aided coppicing, 53, 143–44
machine harvesting
as cost category, 83, 83t
details of picking nuts, 157–160, *158*, *159*
fuel for, 220
Illinois planting, 226–27

planning for, 53
plant spacing for, 67
of seed, 64
selecting for, 9, 15, 58–59, 208
site requirements, 54
machine planting, 53, 91–92, *91*, *92*
maize. *See* corn
marginal care, defined, 78. *See also* care levels
marketing, 177–185
developing markets, 179–183
goals for, 52–53
high-end, 183–85, *184*
non-nutmeat products, 177–79, *178*, 183
principles of, 182
setting prices, 181
marketing order, Oregon and Washington, 42
marking bushes, 145, 152
Marschner Map, 4
mature hazelnuts, 135–150
coppicing, 142–44, *142*
defined, 135
density of planting, 144–48
insect and disease management, 144
nutritional requirements, 17, 135–140, *136*, *137*, *138*
protecting the crop, 149–150, *150*
soil pH, 140–41
soil testing, 136–37, 140
weed control, 148–49, *148*
measuring tapes, 90
meat products, hazelnut-fed, 194
medical-grade charcoal, from hazel shells, 191
Merrell, David, 199, 200
mesh bags, 151, *152*, 156
mice control, 99, 129, *129*, 155, 161–62, *162*
Miller, Greg, 221
mimicry, 34, *34*
mineral fertilizers, 53

Minnesota Department of Agriculture, 29

miscellaneous labor/travel/ expenses cost category, 80, 80–81*t*

mites. *See* big bud mites

modeling, financial, 76–77

modern agriculture, problems with, 23–37
    air pollution, 30–32
    biodiversity loss, 32–35
    drought, 35–36
    flood, 36–37
    overview, 4, 23
    soil loss, 26–28, *27*
    water contamination and loss, 29–30
    woody agriculture solutions, 23–26, *24*

"Mohawking," 97, *97*, 100

moisture content of nuts, 163–64, 169–170

Mollison, Bill, 24

mortality rates. *See* survival rates

mowing
    as cost category, 79, 80–81*t*, 82, 82–83*t*
    for pest control, 125, 127, 128, 129, 133, 149, 155
    year 1 requirements, 97, *97*
    year 2 requirements, 100
    years 3 and 4 requirements, 101

mulching
    during drought, 102
    example planting plan, 74–75
    year 1 requirements, 97, *98*

National Commodity Crop Productivity Index, 56

Native Americans, crop development, 2, 201–2, 219

neglect, potential benefits of, 87

neohybrid, as term, 199

neohybrid hazelnuts. *See also* hybrid hazelnuts

advantages of, 5, 15

air pollution mitigation potential, 31–32

biodiversity restoration potential, 33–35

climate adaptations, 55–56, 214

coppicing qualities, 7–8, 218–19

crown tissue, *9*

current status of development, 17–21

drought tolerance, 35–36

early Badgersett Research Farm work, 3–6

EFB resistance, 5, 107, 108, 109, 205, 207, 208

flood tolerance, 36–37

genetic diversity of, 12, 20, 118, *153*, 197, 201–6, *203*, 214–18, *217*

hybrid hazelnuts vs., 11–12, 199

machine harvesting characteristics, 9, 15

origin of, 10–15

productivity potential, *12*, 15, 17

seed-grown vs. clonal supply, 57

soil restoration potential, 27–28, *28*

water protection potential, 29–30

worldwide crop vs., 39–40

nettle control, 149

New York State, hazelnut production, 57, 185, 248–49

New Zealand, hazelnut production, 44

nightshade weed control, 149

nitrogen
    adapting corn recommendations to hazelnuts, 135–36
    fertilizers for, 29–30, 139–140

importance of, 135

visual assessment of plant nutrient status, 137–38, *137*

no care, defined, 78. *See also* care levels

non-food products. *See* co-products

non-mowing weed control, 79, 80–81*t*

Northern Nut Growers Association, 5, 25, 246

north-south vs. east-west row placement, 67

NRCS web soil survey, 56

nurseries, 247–49

nutcrackers, 151, 174–75, *174*, 227

Nutella, 41, 44, 184

nut growers associations, 246

nutritional requirements. *See* fertility; fertilizers

nuts
    American hazels, 13
    beaked hazels, 14
    in example plan summary, 80, 80–81*t*
    income from, 80–81*t*, 81
    neohybrid clusters, 18, 19, *19*
    removing from young plants, 92–93, 101

nutshells. *See* shells

nutty, as selling term, 180

nut weevils, 114–16, *114*, *115*, 207

oak-hazel savanna landscapes, 7

Oikos Tree Crops, 248

oil palm, 26, 199

Ontario, Canada, hazelnut production, 185

Oregon, hazelnut production, 33, 39, 40, 42–44, *42*, 183

Oregon State University, 43

organic fertilizers, 52–53, 139–140

organic production, 52–53

orientation of rows, 67–68, 72
Oxbo Korvan blueberry-picking
  machines, 157

Pacific Northwest hazelnut
  production
    market for, 183–84
    Oregon, 33, 39, 40, 42–44,
      42, 183
    organic certification, 52
    susceptibility to EFB, 107
    use of clonal hazels, 57
    Washington State, 42–44,
      43–44
Pacific yew, 33
paint, ground marking, 90
paradigm change, 32, 37, 48–49,
  49t
pastured poultry. *See* poultry
  raising with hazelnut culture
patience, as requirement, 50
pellets, hazelwood, 188
pellicles, flavor and, 170, 171
perennial agriculture. *See* woody
  agriculture
perithecia (fungus fruiting
  structures), 109
permaculture, 24. *See also* woody
  agriculture
pesticides
    air pollution from, 31
    barriers to drift from, 65, *65*,
      72–73
    biodiversity loss from, 32, 33
    neohybrid lack of need for,
      32, 220
pest management, 105–33. *See
  also* diseases; ecosystem pest
  management
    aphids, 122
    big bud mites, 116–18, *116*,
      117
    birds, 130–32, *130*, 150,
      155
    black bears, 133
    caterpillars, 120–21, *121*

chipmunks, 127–28, *127*,
  149, 155
deer, 97, 99, 101–2, 132, 155
ecosystem complexity
  benefits, 32–33, 35, 105–6
flea beetles, 120, *120*
grasshoppers, 121–22, *122*
husking before transport for,
  158
Japanese beetles, 118–19,
  119
for mature plantings, 144
mice, 99, 129, *129*, 155,
  161–62, *162*
nut weevils, 114–16, *114*,
  115
organic methods, 52
overview, 105–7
perennial vs. annual
  agriculture, 48–49
pocket gophers, 79, 80–81*t*,
  99, 100–101, 123–26, *123*,
  150
rabbits and hares, 97, 99,
  133
raccoons, 99, 133
site requirements and, 54
squirrels, 128–29, 149, 155
stem borers, 111–13, *112*,
  113, 208
thirteen-line ground
  squirrels, 126–27
for tubelings, 71–72
wildlife attraction to
  hazelnuts, 63–64, *63*
woodchucks, 101, 130, 160
pharmaceuticals, from husks,
  193–94, *193*
Phillips, Michael, 106
pH of soil, 56
phosphorus fertilizers, 135–36,
  139
picking bags, 151, 156
pick-your-own operations, 58
planting, 88–93. *See also*
  preparing to plant
    checklist for, 89*t*

as cost category, 79, 80–81*t*,
  82, 82–83*t*
depth for, 90–91, *93*
follow-up steps, 92–93
machine planting, 53, 91–92,
  91, *92*
receiving and holding
  plants, 88
root care during, 89–90, 92
scheduling, 60, 88–89
tools for, 89, 90
plant spacing, 65–68, *66*, *67*, 73,
  144–46
plant vigor, 109, 113, 135–144
pocket gopher control, *123*
    as cost category, 79, 80–81*t*
    establishment period, 99,
      100–101, 123
    mature plantings, 150
    overview, 123–26
pollination, 8
Polyphemus caterpillars, *121*
poplar, whole crop carbon
  measurement, 216
pork, hazelnut-fed, 194
Porter, Paul, 194
Portugal, hazelnut production,
  44
posthole diggers, 89
potassium fertilizers, 94,
  135–36, 139
poultry raising with hazelnut
  culture
    insect control with, 116, 122
    manure benefits, 141
    overview, 51–52, 100
    soil pH benefits, 116
    sub-food-grade nuts as feed,
      194
predator kites, 126, *126*, 131,
  132, 150, 155
predator support
    for bird control, 131, 132
    as cost category, 79, 80–81*t*
    for rodent control, 125–26,
      127, 128, 129, 149
preparing to plant, 47–85

cost calculations, 75–85, 77*t*, 79*t*, 80–81*t*, 82–83*t*, *84*
decision to plant, 47–50, 49*t*
example planting plan, 72–75, *72, 74*
experimentation considerations, 68–69, *70*
field layout, 64–68, *65, 66, 67*
goals for planting, 50–53, *51*, 54
ground preparation, 56, 69–70, *70*
site requirements, 49, 54–56
tubeling preparation, 70–72
what to plant, 56–64, 58*t*, *61, 63*
pricing of hazelnuts, 45, 181
probably, as word to avoid, 88
processing hazelnuts, 163–176
cleaning husked, in-shell nuts, 165–67, *166*
cracking nuts, 174–76, *174, 175*
husking nuts, 83, 83*t*, 163–65, *164*
quality control, 168–174, *171, 172*
roasting nuts, 176
sizing nuts, 167–68, *167*
produce bags, 151, *152*
productivity of plants
coppicing enhancement of, 66
National Commodity Crop Productivity Index, 56
neohybrids vs. traditional, 15, 17, 18, 19, *19*
selecting for, 207
propagation, in-situ, 146–48
pruning, 107. *See also* coppicing
purse-nurse crops, 76, 80, 80–81*t*, 140
pustules, Eastern filbert blight, *108*

Quackenbush, Henry, 174

quality control
grading nuts, 172–74
judging taste of nuts, 168–172, *171, 172*
quarantine on imported hazels, Pacific Northwest, 43–44
quarterstaffs of hazelwood, 189
questions to ask before planting, 48–50
quicklime application, 141

rabbit control, 97, 99, 133
raccoon control, 99, 133
Radulski, Griff, *96*
rain requirements, 56, 93. *See also* watering
rakes, 90, 92
rancidity of nuts, avoiding, 170, 172
raptor roosts
for bird control, 131, 132, 150
in example planting plan, 73, *74*
field layout considerations, 64
overview, 74
for rabbit and hare control, 99, 133
for rodent control, 129, 149
raschel knit mesh bags, *152*
Red Fern Farm, 248
red squirrels, 128
regulatory concerns, 42, 43, 180
replanting
avoiding need for, 21
considerations for, 65, 75
year 2 checks, 100
residual chemicals, 103
resilience, need for, 211–13, 219
resources, 243–49
ripening of nuts
dates of, 58
observations of, 151, 152–55, *153, 154*
roasting nuts, 176

rodent control
chipmunks, 127–28, *127*, 149, 155
mice, 99, 129, *129*, 155, 161–62, *162*
pocket gophers, 79, 80–81*t*, 99, 100–101, 123–26, *123*, 150
protecting seed from, 146
soil drains by, 30, 220
squirrels, 128–29, 149, 155
storage of hazelnuts and, 161–62, *162*
thirteen-line ground squirrels, 126–27
woodchucks, 101, 130, 160
Rogers, Lynn, 133
rolling, in ground preparation, 69–70
roots
benefits of low care, 87
breeding considerations, 8
care during planting, 89–90, 92
extensive nature of, *10*
plain vs. storage, *9*
pocket gopher threat, 123, 124
soil restoration role, 27–28
rotational cattle grazing, 52
rows to plant, 78, 79*t*
Russia, hazelnut production, 44
Rutter, Elly, *8*

safety concerns
food safety, 175–76, 180
during harvesting, 160
salinization, 29
sampling graded nuts, 172–73
sanitizing shells of nuts, 175–76
sawfly larvae, *121*
scale of production, 223, 228–230
scarifying sides of planting holes, 69
scheduling. *See* timing

science, as tool, 230–31. *See also* experimentation

screening nuts, 166, *167*, 168

seed
  direct seeding, 63–64, *63*
  hand-collecting, 146
  hybrid, 20–21
  saving, 20–21
  seedling types, 59–63, *61*, *63*

seed-grown vs. clonal hazels. *See* clonal vs. seed-grown hazels

selection cycle vs. generation, 207

semi-controlled crossing, 207

separating husks and nuts, 165–67, *166*

settling of soil, 91

shade, avoiding acclimation to, 71, 88

shaker tables, 166, *166*

Shaw, George Bernard, 231

sheep raising with hazelnut culture
  multiple benefits of, 139–140, 149–150, *150*
  planning for, 52
  symbiotic relationship of, 198–99

shells
  co-products from, 190–92
  interpreting markings on, 170–71, *171*
  selling, 177–79
  weight percentage of nut from, 190

Sherman, Clay, 25

shooting pests, 125, 130, 133

shovels, 89, 90

sickle bar mowing, 149

simazine herbicide, 97

site requirements
  field size, 75–76
  land tenure considerations, 49
  pre-planting considerations, 54–56

size of nuts, 58, 207

sizing nuts, 167–68, *167*

Slate, George, 5

slope of land, 54

slowness and sustainability, 50, 75, 85

Smith, J. Russell, 24–25

smoke wood use of hazelwood, 189–190

snow traction, with hazel shells, 192

soil. *See also* ground preparation
  fertility of, 56, 191
  loss of, 26–28, *27*
  settling of, 91
  testing of, 136–37, 140

soil pH
  modifying, 116, 140–41
  neohybrid requirements, 56
  weevil level correlation, 115, 116

South Africa, hazelnut production, 44

soybean industry, as model for hazelnuts, 181–83, 220–21, 229–230

spacing of plants. *See* plant spacing

Spain, hazelnut production, 44

specialty markets, 179–180, 183–85, *184*, 194

speciation, hybrid, 200–201

spider camouflage, 34, *34*

spoilage concerns, 37, 161, 171, 172

spores, Eastern filbert blight, *108*, 109

square-weave bags, *152*

squirrel control, 128–29, 149, 155

state of the crop
  geographic distribution, 225–27, *226*
  overview, 225–28
  role of farmers, 228–230
  tools for change, 230–32

steep slopes, 54

stem borers, 111–13, *112*, *113*, 208

stems
  American hazels, 13
  breeding considerations, 8
  European hazels, 15
  lifespan of, 6

stigmas, 9, *11*

storage of hazelnuts
  flavor considerations, 169–170
  hand-collected seed, 146
  overview, 161–62, *162*
  rodent control, 129, 161–62, *162*
  spoilage resistance, 37

storage roots, 9

storm resistance of hazels, 219–220

strip tillage, 69, *70*

sub-food-grade nut uses, 194–95, *195*

subsoiling during ground preparation, 56, 69

sunlight distribution, 215–16

sunny conditions, acclimating tubelings to, 88

survival rates
  benefits of neglect, 87
  care levels and, 88
  first year, 66
  replanting decisions and, 100

sustainability and slowness, 50, 75, 85

Swanson, Bert, 29–30

Switzerland, hazelnut production, 44

symbionts, 198–99

synthetic fertilizers, 53, 136–37, 139–140

target products and markets, 52–53

taste of nuts. *See* flavor of nuts

teosinte (corn precursor), 2, 19, 201–2, *202*, 228

testing of clonal hazels, 57

test plantings, 47

theft of nuts, 151, 155

thirteen-line ground squirrels, 126–27

thistle control, 149

tilling
neohybrid lack of need for, 220
in preparation for planting, 69–70, *70*

Timeless Tools, 247

timing
coppicing, 142–43, 187
dividing plants, 147
egg spray, 132
fall coloration, 138
fertilizers, 137
germination of hand-collected seed, 146
harvesting, 152–55, *153, 154*
planting, 60, 88–89
removing nuts from young plants, 92–93, 101
ripening of nuts, 154

tools and equipment
as cost category, 82, 82–83*t*
for harvesting, 151–52
husking machines, 163, 164–65, *164*
nutcrackers, 151, 174–75, *174*
for planting, 89, 90
resources, 246–47

total yearly expenses cost category, 80, 80–81*t*

tractor costs, 81, 82–83*t*

traditional hybrid hazelnuts. *See* hybrid hazelnuts

transport of hazelnuts, 158, 161–62, *162*

trapping pests, 124–25, 133, 150

trazels (Turkish tree hazel-hazelnut cross), 12

tree crops. *See* woody agriculture

Tree Crops: A Permanent Agriculture (Smith), 24–25

tree growth form
bush form vs., 6, 188

European hazels, 15
Oregon production, 42, *42*

tree shelters/tree tubes, 98, 99

truffle growing potential, 192

tubelings
care during planting, 90–91
flooding vulnerability, 102
infilling with, 146
overview, 60–62, *61*
planting checklist, 89*t*
preparing for planting, 70–72
removing nut from, 92–93

Tunisia, hazelnut production, 44

Turkey, hazelnut production, 3, 39, 40–41, *41*, 52, 185

turkey raising with hazelnut culture, 51–52

Turkish hazels, 40

Turkish tree hazel-hazelnut cross (trazels), 12

twig burnout, 136, *136*

Twin Ponds, 248

Ukraine, hazelnut production, 44

unfair advantage concept, 180

United Nations Food and Agriculture Organization (FAO), 1, 39

United Nations International Fund for Agricultural Development, 29

University of California at Berkeley, 215

University of Minnesota, 17, 29–30, 126, 225

urea, feed grade, as fertilizer, 94

USDA hardiness zones, 55

US Department of Energy, 73

value-added nut products, 179–180

velvetleaf cover crops, 98

ventilation during transport, 161

Vexar produce bags, *152*

visual assessment of plant nutrient status, 96, 137–38, *137, 138*

volume price breaks, 82

Walser-Kolar, Denise, 11

Washington State, hazelnut production, 42–44, 43–44

water contamination and loss, 29–30

watering
as cost category, 79, 80–81*t*, 82, 82–83*t*
during drought, 79, 94, 102
immediately after planting, 92
rain requirements, 56, 93
tubelings, 71
year 1 requirements, 93–94
year 2 requirements, 100

wattles of hazelwood, *178, 189*, 218

weather tolerance of hazels, 221

weed control. *See also* mowing
as cost category, 82, 82–83*t*
during harvesting, 151
husking before transport concerns, 158
for mature plantings, 148–150, *148, 150*
non-mowing weed control, 79, 80–81*t*
woody weeds, 83, 83*t*, 148–49, *148*
year 1 requirements, 96–98, *97, 98, 99*
year 2 requirements, 100
years 3 and 4 requirements, 101

weevils. *See* nut weevils

Weschcke, Carl, 5

white-footed mice control. *See* mice control

white-tailed deer control. *See* deer control

whole crop carbon measurement, 216–17, *217*

wick herbicide application, 97

wild cucumber control, 148–49, *148*

wildlife habitat creation, 54, 58, 92–93

wild parsnip control, 149

wilting, dangers of, 94

wind protection, 55, 73–74

wire cages, 101, *101*

wire stake flags, 90, 101, 152

wood ashes, for adjusting soil pH, 141

woodchuck control, 101, 130, 160

wood harvesting
   coppicing schedule considerations, 142–43
   income from, 83, *178*, 179, 187

types of co-products, 187–190, *189*

woodstoves, hazel products for, 190

woody agriculture. *See also specific crops*
   annual agriculture vs., 2–3, 48–49, *49t*, 221
   Badgersett Research Farm short course, 203, 245–46
   benefits of, 3, 23–26, *24*
   biomass production potential, 31–32, 216–17, *217*
   book resources, 245
   domestication of tree crops, 25–26
   mindset required for, 48–49, *49t*

photosynthetic activity vs. annual crops, 217–18

woody weeds, 83, 83*t*, 148–49, *148*

world hazelnut industry, 39–45
   Italy, 39, 41
   opportunities in, 3, 45
   other production areas, 44–45
   overview, 39–40
   Turkey, 3, 39, 40–41, *41*, 52, 185
   United States, 42–44, *42*

Zea diploperennis, 3. *See also* teosinte (corn precursor)

zones of hazels, 64

Z's Nutty Ridge LLC, 248–49

green press INITIATIVE

Chelsea Green Publishing is committed to preserving ancient forests and natural resources. We elected to print this title on paper containing at least 10% postconsumer recycled paper, processed chlorine-free. As a result, for this printing, we have saved:

14 Trees (40' tall and 6-8" diameter)
6,726 Gallons of Wastewater
219 million BTUs Total Energy
450 Pounds of Solid Waste
1,240 Pounds of Greenhouse Gases

Chelsea Green Publishing made this paper choice because we are a member of the Green Press Initiative, a nonprofit program dedicated to supporting authors, publishers, and suppliers in their efforts to reduce their use of fiber obtained from endangered forests. For more information, visit www.greenpressinitiative.org.

Environmental impact estimates were made using the Environmental Defense Paper Calculator. For more information visit: www.papercalculator.org.

About the Authors

PHILIP RUTTER is the chief scientist, founder, and CEO of Badgersett Research Farm, founding president of The American Chestnut Foundation, and past president of the Northern Nut Growers Association. He is an evolutionary ecologist, with a master's and "ABD" (All But Dissertation of PhD) in zoology, with a minor in animal behavior. At one point he escaped from academia, when he discovered it was not his cup of cappuccino. With a parasitologist PhD advisor, he is deeply trained in the evolution of diseases and symbiotic systems.

BRANDON RUTTER-DAYWATER

DR. SUSAN WIEGREFE is Badgersett's research associate. She has a PhD in plant breeding and plant genetics and taught courses in plant propagation and nursery management for four years at the University of Wisconsin–River Falls. Co-incorporator and past president of the North American branch of The Maple Society, her latest personal endeavor is as the owner and operator of Prairie Plum Farm, where she raises Babydoll sheep, fruit, and nuts, and soon will include an aquaponic vegetable/tilapia system. In her spare time she hangs out with her two Havanese dogs, when she's not spinning or making cheese and beer.

SABINA PETERS-DAYWATER

DR. BRANDON RUTTER-DAYWATER grew up on Badgersett Farm, eating some dirt but very few hazelnuts—they were all for seed! Dedicated to the long-term viability of the human race, and therefore our concomitant living things, his formal training is primarily in engineering and biologically inspired robotics. A national merit scholar upon graduating from high school, now he's the COO at Badgersett, building a family and a house where he's convinced he'll be able to do the most good. He is now growing and eating a lot more hazelnuts!

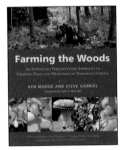

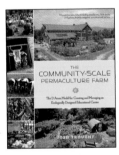